CAMBRIDGE COUNTY GEOGRAPHIES

General Editor: F. H. H. GUILLEMARD, M.A., M.D.

THE

NORTH RIDING

OF

YORKSHIRE

T0349125

Cambridge County Geographies

THE
NORTH RIDING
OF
YORKSHIRE

by

Capt. W. J. WESTON, M.A., B.Sc.

With Maps, Diagrams and Illustrations

Cambridge :
at the University Press
1919

CAMBRIDGE UNIVERSITY PRESS
Cambridge, New York, Melbourne, Madrid, Cape Town,
Singapore, São Paulo, Delhi, Mexico City

Cambridge University Press
The Edinburgh Building, Cambridge CB2 8RU, UK

Published in the United States of America by Cambridge University Press, New York

www.cambridge.org
Information on this title: www.cambridge.org/9781107622449

© Cambridge University Press 1919

First published 1919
First paperback edition 2013

A catalogue record for this publication is available from the British Library

ISBN 978-1-107-62244-9 Paperback

CONTENTS

ILLUSTRATIONS

ILLUSTRATIONS

vii

MAPS

The illustrations on pp. 12, 67, 102, 118, 119, 145 are from photographs kindly supplied by the Rev. W. Denison, that on p. 52 is reproduced by courtesy of the Tees Conservancy Commission; those on pp. 72, 74, 78 by permission of Messrs Bolckow, Vaughan and Co., Ltd., from photographs supplied by the Photochrom Co., Ltd., that on p. 120 by permission of Mr C. F. Innocent and Mr J. Kenworthy, that on p. 143 is from a photograph kindly supplied by Sir Hugh Bell, Lord-Lieutenant of the County.

The photographs reproduced on pp. 3, 8, 24, 25, 27, 32, 34, 42, 59, 65, 107, 112, 123 were supplied by Lieut. A. H. Robinson, those on pp. 5, 7, 11, 15, 17, 18, 20, 23, 28, 49, 81, 93, 108, 113, 117, 122, 124, 127, 129, 130, 135, 147, 151, 152 by Messrs Frith and Co., Ltd., those on pp. 46, 68, 98, 157 by Mr E. Hall, those on pp. 139 and 141 by Messrs Mansell and Co., Ltd., and that on p. 110 by Mr F. H. Crossley.

1. County and Shire. The names *Riding* and *Yorkshire*; their Origin and Meaning.

Yorkshire is of so great extent that it is not surprising to find that it has from very early times been treated as three separate counties. Each Riding has its own Lord-Lieutenant and its own county offices; and, though the City of York is, for some purposes, the capital of the whole, each has its own county town. That for the North Riding is Northallerton.

The English kingdoms that were founded in our island after the departure of the Romans could not, in those days of difficult travelling, be properly governed from one point. They were therefore divided into *shares* or *shires*, roughly according to the number of men able to fight; and the men of these shires were themselves answerable for most duties of government. Yorkshire was a portion *shorn* off (the derivation of the word is the same as that of shire) by the wedge of "Bishop's land," the county of Durham, from the great kingdom of Northumbria.

The shire was ruled for the king by an officer, usually called an Ealdorman; the Lord-Lieutenant of to-day is his representative. When the Normans came they called him *Comte* or Count; and the district over which he had control was a *Comitatus*, a *Comté*, or County. But a good many of the shires did not become counties; the Norman Count would often rule over several shires. Thus, even as late as the reign of Elizabeth, the traveller Camden speaks of the North Riding as made up of four shires. These were Richmondshire, Swaledale and the adjoining plain given by the Conqueror to his cousin; Clevelandshire, the land of cliffs; Allertonshire; and Blackamore, the black moorlands, the old name for the bleak moors between Cleveland and Pickering Vale. The name Richmondshire is still used to denote one of the four divisions of the county having a member in parliament.

A shire usually took its name from its chief town. This is the case with Yorkshire, of which the Anglian name was *Eofervice-scyre*, the shire of York. The Anglian city was called *Eofervice*, the "over-city." When the Danes ruled in York they gave the Anglian name a Danish sound, *Jorvik* (the *j* being pronounced as *y*), and in course of time this word of two syllables becomes the modern York. Or, as some think, *Jorvik* may mean "the settlement (*vik*) on the Ure" and have no relation to the Anglian name.

The name "Riding" also is due to the Danes, who, as we shall see, had much to do with Yorkshire. The word has lost its first sound *th* because it was so often preceded by the words North, East, or West. North-riding is really North-thriding. The *th* would be the more easily lost as the two words were mostly written as one. For instance, we have in the Rolls of Parliament for 1474, "The Shire of York in the Estrithing, North-rithyng, and Westrithyng of the same." *Thri* is kin to the

Across the Yorkshire Moors

Danish for *three*, and *ding* or *thing* is the part or shire that has an assembly of its own. We meet with this syllable *ding* or *thing* or *thring* wherever the Danes have settled: thus in the North Riding, far up among the mountains of the west, is the little village of **Thringarth**, the garth or meadow in which men gathered to talk

things over. A Riding, therefore, is a third part of a
larger whole, which part nevertheless has an assembly—
a shire-moot—of its own.

York was known in Roman times as Eboracum,
and the first two syllables of this name are still in use.
The Archbishops of York employ them as the official signa-
ture, preceded by the Christian name, as Cosmo Ebor.

2. General Characteristics.

The North Riding, though one only of the three
parts into which Yorkshire has long been divided, is
yet fourth in size of English counties, and its wide area
contains much of the very greatest interest and importance. On the west, from beneath the New Red Sandstone and the gravel of the plain, rises the region of
mountain limestone. There, the huddled masses of
mountains, with their secluded valleys, their rapids, and
their waterfalls, afford rugged and grand scenery not
easily equalled. In the Tees Valley also there is striking
scenery. For there the hard igneous rock of the Whin
Sill crosses the valley, and, yielding less easily to erosion
than the softer Carboniferous beds over which it passes,
forms ridges. Across these ridges the river dashes in
imposing falls. Wensleydale, where Aysgarth Force is
one among numberless fine scenes, is with justice called
the Piedmont of England. From Richmond Castle is a
view eastward over a wide stretch of level farmed land;
and Pickering Castle, too, looks over a productive

Bainbridge, the Roman Virosidum

saucer-like vale scooped out long ago by moving ice-fields. The fringe of cliffs that faces the North Sea is broken by many beautiful bays in which nestle a number of towns well placed for fishing and pleasant as holiday resorts. Along the south bank of the Tees, as we approach the sea, we find a crowded district of modern industry, of blast furnaces that turn the ironstone of the Cleveland Hills into material for the iron-worker, of steel-mills, of ship-yards, and of busy docks and quays.

Nine-tenths of the riding is still occupied by farms. The broad valley between the Pennine fells and the North Yorkshire moors on the east is covered by excellent soil. The green fertile plain, spread over the soft sand-stones and marls of the New Red series of rocks, stretches to the sterile moorlands east and west where the lime-stone emerges from its deep foundation. The Yorkshire farmer has long been famed for his skill and energy; York has one of the largest cattle-markets in the country; the Yorkshire breed of horses attracts buyers to the Northallerton fairs from all parts of the Continent; and Yorkshire hams and bacons are well known. But, as in most parts of Britain, the farm-worker is vanishing. At the last numbering of the people, though the population of the whole riding showed during ten years an increase of over eleven in a hundred, every division or wapentake, except two, showed a decrease. The two exceptions were those we may call urban; the rural wapentakes all showed a decrease in the number of inhabitants.

Middlesbrough and Scarborough with their attendant towns had grown rapidly; many a country worker had found employment in the towns or had emigrated. Yet, in spite of these two small areas where people are closely massed, the rest of the county contains so few inhabitants that the North Riding is among the most sparsely peopled of our counties.

Corporation Road, Middlesbrough

The mushroom town, Middlesbrough,. is indeed a remarkable illustration of the manner in which the working of the huge stores of iron ore in the Cleveland Hills is transforming this part of the county—turning it from a quiet farming region into a noisy and bustling manufacturing centre. A hundred years ago the site of Middlesbrough was occupied by a few farmsteads;

now a population of over a hundred thousand finds
work in the smelting and working of iron, and in the
shipping, ship-building, and other industries that iron
has drawn to the spot.

The coast towns—chiefly Scarborough, which is also
the greatest of Yorkshire watering places, and Whitby,

Scarborough, South Bay

where a busy new town has grown up alongside the
ancient one—have a great fishing industry. Scarborough stands fourth of the north-eastern fishing ports,
only Hull, Hartlepool, and Shields having greater
quantities of fish brought in.

Rich as the North Riding is in natural beauty, the

beauty of broken mountains and rushing torrents, of smiling farm lands and of cliff-bordered coast; rich as it is in interest as an example of modern industry, it is richer still in what appeals most to the student of plant and animal life. Upper Teesdale and the wide moorlands are productive hunting-grounds for the botanist; among the dales running into the Pennines certain birds and animals, elsewhere disappearing from our islands, still find refuge; and the varied coast, with its rock-sheltered bays, has a wealth of marine life.

For the historian, too, the county is of special interest. From the earliest time the North Riding must have been an important district. A surprising number of remains of primitive man have been found here, mainly in barrows on the moors. Roman relics—as we might expect seeing that York, the Roman capital of the island, could be called "altera Roma," "a second Rome"—are most plentiful. Whitby Abbey and Stamford Bridge remind us of notable events. Northallerton is only one of the many places on the Great North Road where Englishman and Scot fought. And if in modern times our history is mainly the story of peaceful industrial development, it is not the less but rather the more noteworthy. Nowhere else can we see so clearly the working of what is called the "Industrial Revolution," the change from a farming England to an England of factories and workshops.

The North Riding, which we now proceed to study in detail, occupies that part of the eastern slope from the Pennines which lies south of Durham and north of the

East and West Ridings. But, as we shall see, the boundaries of the counties on the sides of the great divide are not quite along the line of water-parting. The Greta, for instance, flows into the North Riding from Westmorland, where it rises on the flattened height called Stainmore Forest.

3. Size. Shape. Boundaries.

The North Riding is fourth in area of the counties of England. It extends over 1,357,433 acres, or 2121 square miles of land: its neighbour to the south, the West Riding (largest of all our counties), Lincolnshire, and Devonshire, alone have a greater extent. It is a little more than twice the size of Durham, its neighbour to the north; and no fewer than fourteen Rutlands would be needed to cover its surface.

The county forms an irregular four-sided figure with one great angle of land, having York at the point, towards the south. It is roughly an oblong, stretching about twice as far east and west as it does north and south, and having a triangle on its southern side. The longest line that can be drawn within the area is from the north-western corner, where Maize Beck joins the infant Tees, to the south-eastern corner, a little north of where the spur of rock, Filey Brigg, runs a mile into the sea. This line measures about 87 miles, and a line from Redcar to York, a distance of 45 miles, is the longest that can be drawn directly north and south.

The very irregular outline of the riding is due to the nature of its boundaries, which on east, north, and west, are respectively sea, river, and mountain. They

Hell Ghyll Ravine, on the Westmorland border

are what we term natural boundaries. On the fourth side, too, the line marking off the North from the East and West Ridings is mainly one of mountain ridge or of river course.

The eastern boundary is the North Sea, which curves round from a point north of Filey Brigg to Tees Mouth. The seaward edge of the Yorkshire heights fronts the

The Ouse at York

waves with a bold stretch of high cliffs. The Tees throughout the greater part of its length, tearing its way through the mountain rocks, or slowly crossing the plain at Middlesbrough, separates the riding from Durham, and forms the northern boundary. The western border extends from where Maize Beck flows along the flanks of Mickle Fell to join the Tees to Newby Head, where at a height of 1400 feet the road from Westmorland climbs into Yorkshire. This part of the border follows roughly the line of greatest height in the Pennines. For a short distance where it runs close to and parallel with the infant Ure, the Eden forms the county boundary, running in a dark limestone ravine. We shall have opportunity for studying these three boundaries in later chapters.

From Newby Head the boundary line stretches towards York to the south-east. It is formed by the crest of that spur from the Pennines which separates Wensleydale in the North Riding from Wharfedale and Nidderdale in the West Riding. The summit of Buckden Pike, 2302 feet high, one of the few *pics* or pointed mountains, is cut by the boundary. Farther east on the border are Great Whernside and Little Whernside. This crest reaches the river Ure at Hack Fall, a deep wooded ravine through which the Ure rushes swiftly. The river, called the Ouse after it receives the Swale, then forms the boundary up to York, though in parts near Ripon and Boroughbridge the West Riding extends over to the north side of the stream.

From The Wyke north of Filey Brigg the border line crosses the coast range of wolds to the river Derwent. Two miles from the coast the river Hertford flowing west to the Derwent along the low-lying "carrs" is the border, and continues so to its junction with the main stream. The border then follows the wanderings of the

Derwent south-westwards towards York, past Malton to Stamford Bridge. Here the Derwent makes a sudden turn to the south, and the boundary, artificial for about ten miles, crosses the plain past Holtby till we reach York. The East Riding is the neighbouring county from The Wyke to York; the West Riding from Newby Head to York.

The Ainsty of York, in which the city stands in proud isolation, constitutes a county in itself, and is an instance of the confusion caused by the varying divisions of English counties. For purposes of local government the Ainsty is a wapentake of the West Riding; for parliamentary purposes, however, it forms part of the Thirsk division of the North Riding.

There are now no parts of the North Riding separate from the main area as in former times; nor are portions of other counties contained within its borders. In early days the Bishops of Durham had rule over the old village of Crayke and the adjoining region within three miles. In 1844 Parliament made this survival of the prince-bishopric a part of the riding.

4. Surface and General Features.

The North Riding may be divided into three main areas. The western part, beyond the mound-and-river fort of Richmond, is the region of mountain moorland and rough pasture, adding little to the wealth of the county. Breaking up this region are the wider dales of the Ure, Swale, and Tees. The northern extension of the fertile Vale of York—the Vale of Mowbray it is at times called because of the great nobles who lived and

Saltwick Bay, Whitby

fought there—separates the Pennine fells from the North Yorkshire moors, a huge limestone slab lower than the Pennine ridge, but more bleak and dreary, broken by narrower dales, and less easy to traverse. The one tolerable road southward is that running through Bilsdale from Stokesley to Helmsley. Beyond the moors and stretching to the Derwent is the Vale of Pickering.

The eastern moors abut upon the sea in a line of lofty cliffs—the loftiest indeed along the English coast—and fall rather abruptly to the plain of the lower Tees. Though less interesting in scenery this plain, a portion of Cleveland, is by far the most important part of the county by reason of its proximity to easily-won stores of ironstone, to the coal of Durham, and the lime of Loftus. It is the industrial part, containing half the population.

The county as a whole lies higher above the sea than do most English counties: wide stretches, as will be seen from the map inside the front cover, are above the 1200 feet line, and little is low-lying. The highest summit of all is that of the fitly-named Mickle Fell on the western moors, south of Maize Beck. Here, just within the riding, the millstone grit cap of the mountain is 2591 feet above the sea. It is not a striking peak, and Great Whernside on the West Riding border, though lower (2310 ft high), is more conspicuous and better known. So too is Penhill, rising abruptly in the midst of Wensleydale, and crowned, like other noticeable heights, with its ancient beacon. In the north-east Roseberry Topping rises in a similar abrupt manner from the Cleveland Plain, and is therefore more famous than the much higher Burton Head farther on the plateau.

On the western fells heather and peat cover the millstone grit, for this hard rock affords poor soil. Over these cool heights with their abundant moisture is a wilderness of attractive but nearly useless plants, ling, crowberry, cotton-grass, sphagnum moss, bracken, and heather; and in the damper parts the plants have formed

View from Leyburn Shawl

masses of black peat. When the grit meets the limestone the heather disappears before the vivid green of the lime-loving grass, and mountain pasture succeeds to moorland.

From the central ridge of the Pennines, that uneven block which fills the west of the riding, run spurs to the east. These, narrowing in breadth and decreasing

in height, divide and confine the dales. Separating
Wensleydale from the West Riding a spur passes through
Knoutberry Hill, Whernside, and Penhill to the plain
at Masham. North of the Ure from Great Shunnor
Fell (2546 ft high), between the headwaters of Swale
and Ure, a monotonous ridge descends to the plain near

Roseberry Topping

Ainderby, forming in its course the strange limestone
terrace called Leyburn Shawl.

The greater Yorkshire dales have little communication across the mountain ridges, which do not bend into
convenient passes. The mountain road from Hawes to
Muker, for instance, is forced to climb to a height of
1700 ft. This is Buttertubs Pass, bearing this curious

name from the presence along its course of deep holes where the limestone has been dissolved.

East of Thirsk a steep scarp, the western ridge of the moors, rises from the Vale of York. This ridge is the Hambleton Hills, rising on heather-clad Black Hambleton to 1257 ft above the sea. The Cleveland Hills are the northern part of the eastern moors: they rise in Burton Head, near the source of the Esk, to a height of 1489 ft, and in the cone-shaped Roseberry Topping[1] to 1057 ft. At Eston Nab[1], with its rich ironstone deposits, the plateau drops to the Tees Valley.

The Carboniferous Limestone gives rise to bold scenery; and in what are called the Yoredale rocks (from their having been studied most closely in the valley of the Ure or Yore) the successive steps of shale, sandstone, and limestone form the terraces and steps so frequent in Wensleydale.

The fertile Vale of Pickering is of great interest. An area of 160 square miles is hemmed in by hills on all sides except towards the sea; its floor is a level flat of alluvial scourings brought down by streams. This alluvium covers a deep-seated valley cut out in the clay. The drainage of the whole southern slope of the eastern moorlands gathers into the Derwent. But this deep trench, instead of reaching the sea eastward, is twisted back in a most curious way and cuts through the Howardian Hills to the Vale of York.

The physical feature, however, that most distinguishes

---

[1] In the North Riding a "nab" is the steep cut end of a ridge dropping abruptly to the plain ; a "topping" is a peaked height.

the riding from other English counties is the number, size, and beauty of its dales. The three great dales threaded by the Tees, Swale, and Ure are very attractive, as indeed are their tributary dales, penetrating deep into the mountain ridges, traversed by a brawling beck,

A Dale Town: Muker in Upper Swaledale

and often graced by a ruined abbey or castle. The eastern dales are narrower and wilder, are rather gorges in the moors; but they, too, are delightful, and some, like Forge Valley and Esk Dale, are famous for their beauty.

5. Watershed. Rivers.

The North Riding is, in the main, a part of the longer and more gentle eastern slope from the great arch of rock that we call the Pennine Range; and with the slope towards the North Sea is also a more gradual slope to the northern end of the English Plain. The larger rivers, the Tees, Swale, and Ure, have therefore a general south-easterly direction.

The western boundary of the basins from which the Yorkshire rivers draw water follows roughly the line of greatest elevation of the Pennines. This boundary is the main watershed of the rivers. It separates for instance (on the slope of Lund's Fell they are only a few yards apart) the head-waters of the Ure from those of the Eden. We should note, though, that like other features even of the solid land the watershed is not fixed. It is slowly retreating eastward; more water is being carried to the Irish, less to the North Sea. It is clear for instance—the width of its bed makes it obvious —that the Maize Beck once gave far more water to the Tees than it does now. It has been beheaded by a stream flowing westward.

Two things working together account for this diversion of water from the east. One is the steeper western slope of the Pennines; the other is the greater rainfall on the west. Both these give to the western streams more excavating power, and so cause them to eat back into the mountain ridge.

The eastern spurs from the Pennines separate the drainage areas of the Tees, the Swale, and the Ure. The rivers running through the vales thus formed—Teesdale, Swaledale, and Wensleydale—are like the main veins of a beech leaf, the smaller becks that dash down the enclosing mountains are the tributary veins.

On the huge uneven slab of limestone that almost fills the eastern portion of the riding one cannot define a clear water-parting. The plateau is intersected by deep dales through which peat-stained streams gather southward into the Derwent or into its chief feeder the Rye, or northward into the Esk and the Leven. The Esk is the chief river of Cleveland and the only river of any size that breaks through the cliffs to the sea. At Cronkley Gill it cuts its way by a deep ravine through the moorland ridge.

The rivers of the riding seldom lack a copious supply of water: the rainfall on the lofty and broad moors is great and evenly spread over the year, and the tracts of moss and peat-covered bog store up plentiful reserves. As in other limestone districts there is much water circulation underground. The Greta, for instance, has worn for itself at one point a way through the limestone and has formed a natural bridge (God's Bridge people call it), 16 feet in span and 20 feet wide, and a short distance away it passes for half a mile underground. The water supply of Scarborough is derived entirely from such underground circulation. It is drawn from the Corallian rocks which rest on the Oxford Clay. These water-bearing layers are tapped by deep wells at

Irton and Osgodby, and by adits in the cliff at Cayton Bay. Gormire too, a few miles east of Thirsk, a small expanse of water sometimes called a lake, has no visible outlet; its waters drain away through pervious rock until they lodge on the clay or find their way into the Rye. The surface water rapidly disappears, and there is an

The Greta near Rokeby

abundance of open joints and cavities. The best instance of these is on the mountain road from Hawes to Muker across the watershed between Ure and Swale. Near the summit of the pass are the Buttertubs, already mentioned.

Where there is a millstone grit cap over the limestone the layers of rock have varying powers of resistance to waste (or weathering). The Millstone grit indeed weathers rapidly, but is not soluble in acid water as the limestone is. And where limestone rests, as usually it does, on softer shales, we have a third layer to add to the unequally resisting rocks. There are, as a result, an extraordinary number of waterfalls on the North

High Force, on the Durham border

Yorkshire rivers. They occur not only along the main streams, but on the becks that plunge down the narrow transverse dales, some of them amid scenes of wonderful beauty. Among them, perhaps because of Turner's painting, Aysgarth is best known. Here the Ure, already a full broad stream, pours over a series of limestone shelves, the lowest fall being the finest. High Force, where the Tees issuing from the confined rocky channel through the Falcon Clints falls over a basalt precipice, is the noblest waterfall in Britain. The

Hardraw Scar

upright volcanic rocks based on the horizontal limestone layers have a striking effect. Hardraw Scar, over which a little feeder of the Ure plunges 100 feet, is a limestone layer on the slope of Shunnor Fell. Below the limestone the soft shales have been eaten away, so that the edge of the falls projects far beyond the base.

Except the Tees and the Esk and a few becks that burst through the cliff barrier of the eastern coast, all the rivers of the North Riding find their way into the Ouse, and so to the Humber.

The Tees, of which the right bank only belongs to the riding, affords in its upper course as beautiful river scenery as any in England. Desolate and bare at first, the surrounding country becomes finely wooded lower down, and, till one reaches the busy industrial region, presents a succession of exquisite views. Through the Falcon Clints, a famous hunting ground for botanists, and High Force, past Romaldkirk and Cotherston through Rokeby woods to Piercebridge, is the most interesting stretch. At the ancient town of Yarm it receives the Leven, draining the Cleveland plain. The river by Thornaby and Middlesbrough and South Bank flows through a wonderful industrial region; a small area, but one in which half the population of the riding finds work and wages. In Baldersdale, through which the Balder flows over hard rocks to the Tees, is the huge reservoir, almost a lake, from which Middlesbrough obtains its excellent water.

The Esk, collecting the drainage high on Stockdale Moor, has carved out a deep dale to the sea at Whitby. Through this natural pathway, road and railway climb across the plateau to Stokesley and Yarm, so that for many miles river, road, and railway journey together. The river follows the line of a great dislocation of the rocky layers by which the hills to the north have been depressed; and at Grosmont it flows over the basalt dyke spoken of in the section on Geology.

The headwaters of the Ure, or Yore, are on the western border, quite near the cleft in the limestone carved out by the Eden. After turning sharply round

Hawes, on the Ure

Great Shunnor Fell the river flows through Wensleydale, closely accompanied by road and railway, until, in the plain near Myton, it converges to the Swale. It then takes another name, the Ouse, and forming part of the southern boundary enters the Ainsty of York. The little market town Hawes is the chief in its upper course;

thence, giving an endless variety of fine river scenery, it passes Bainbridge with its Roman camp, Aysgarth, Wensley, diminished in importance from the time when it gave its name to the whole vale, Middleham, where the castle of the King-maker frowned over against that of the Scropes and Masham, famous for its September

Semmer Water

sheep fair.   One of its tributaries, the little river Bain, expands at the foot of flat-topped Addleborough (1564 ft high) into a small lake. This is Semmer Water —a curious name compounded of three words meaning water, *sea*, *mere*, *water*—the only real lake occurring in the riding.

The Swale—the sources of which are where the lofty

High Seat and Nine Standards Rigg block the western end of the narrowed valley—runs through the wildest and most secluded of the great dales to Richmond. After Richmond the river, no longer a brawling mountain stream closely confined by deep ridges of fells, flows languidly through the productive plain, gathering, by means of feeders like the Wiske, the north drainage of the Vale of York. Muker and Reeth among the mountains, Topcliffe and Myton on the plain, are other towns on the river.

The Derwent, which collects mainly by means of its greater feeder the Rye the drainage of a wild moorland tract and of the Vale of Pickering, rises in the eastern moors quite close to the sea. Failing however (or ceasing) to break through the cliff barrier, it flows through the deep and densely-wooded Forge Valley, past Hackness and Ayton, till on the level called the Carrs it turns westward to form the boundary with the East Riding. It receives the Rye, on which historic Helmsley and Rievaulx stand, from the Hambleton Hills, flows by Malton, and at Stamford Bridge leaves the riding before it reaches the Ouse. The river scenery through the Howardian Hills below Malton is very fine. Only a short distance separates Scalby Beck, which enters the sea two miles north of Scarborough, from the Upper Derwent. A "new cut," carrying off the surplus waters of the Derwent in time of flood, has been made to the beck. Geologists tell us that North Sea ice once overrode the eastern coast. There are clear proofs that at one time an ice wall closed the exit from the Vale of

Pickering eastward. The drainage from the moors therefore filled the Vale and had to find passage westward to the Ouse. The gap at Malton through which the Derwent flows was thus made.

6. Geology.

Study of the geology of a district should teach us how the land has been built and of what material it is made. It should explain to some extent how the lofty fells in the west came into being, and why, half a mile above sea-level, is found limestone that was formed from shells dropped upon the floor of an ancient sea. It should show us why coal is absent and iron present in the county; and we should learn from it not a little concerning the soil, for this is dependent upon the rocks below from which it has been formed. Lastly we should learn something of what took place long before the period of man's advent on the earth.

A visit to the cliff bulwark of the county, or to one of the many ledges (at Aysgarth, for instance) over which the rivers tumble, shows us that the rocks forming the structure of the county are arranged, not in a disorderly heap, but in beds or layers. These layers the geologist calls strata. They are not often found in a horizontal position, though at the time of their formation they usually had a level surface. Slow or sudden movements of the earth's crust have often tilted the layers: thus, in the North Riding the layers bend towards the south, so that, as we travel north along the coast to Redcar, the rocks that are spread to view are ever older. Beginning with the chalk cliffs at Flamborough we meet the earlier series till we come to the Lower Lias, to the shales and sandstones at Redcar.

How the changes have come about we may understand by noticing how a drying and therefore shrinking apple becomes covered with wrinkles. The earth itself is cooling and shrinking, and so pressure is being exerted upon the rocky crust, forcing it into folds as one might a tablecloth. The Pennine Range is the summit of one great wrinkle formed by pressure from east to west: the North Riding is part of the eastern portion of this great arch or dome, or *anticline* as it is called.

Sometimes the pressure was so great that the strata were not only folded up or down, but were actually broken across so that the layers were no longer continuous The greatest of these cracks—or *faults* as they are called—in our county is along the Pennine ridge. Here the layers on the west have been pushed many thousand feet above those on the east. Over the Westmorland border the rocks belong to the division called in the table Silurian—so termed from the name of the British tribe inhabiting the Welsh district where the division appears most; in the North Riding they belong to the Lower Carboniferous, or coal-bearing, division. We may consider the riding as a huge block fault sunk from the rocks adjoining on the west.

Another fault occurs at the mouth of the Esk. This Eskdale fault has raised 200 feet the layers south of

Whitby, where one may easily examine the different
layers in the cliffs; the east pier is built on alum shale—
hard Upper Lias that fronts the sea with a rugged scar
—the west pier is on sandstone. Another and a very
interesting fault is at Staithes. The cliffs on opposite
sides of the little harbour show beautiful but quite

Cavities in Buttertubs Pass

different sections of strata. The signal cliff, Penny
Nab, on the east, has under its later covering hard shale
and ironstone; in Colburn Nab, on the west, we find
beds of loosely compacted sandstone containing many
fossils.

The rocks of the Pennine slope belong, as we have

said, to the Carboniferous (or coal-bearing) division. Only in one small region are there rocks earlier than these.

Three main layers make up this series. Lowest is the Mountain limestone, which forms the core of the western ridge. Covering the limestone, and deposited upon it as grains worn away from hard volcanic rocks, is the Millstone grit. Covering the grit are the coal measures,—the clays and sandstones that contain seams of coal, most valuable of all fossils. But there are no layers of coal in the North Riding, though north of Tees and south of Ure rich deposits are found. It is clear therefore that besides the pressure east and west, the pressure that raised the crest of the Pennines, there must have been in our county an added pressure from north to south. This pressure caused the rocks to curve in an immense arch from Durham to the West Riding. As the ground rose, coming under the influence of the rains and frost above, the arch was slowly pared down: now, just as there is no coal on the mountain tops, where indeed the Mountain limestone itself has in many places been laid bare, so in the same way the North Riding has been robbed of what adds so greatly to the wealth of the adjoining counties. The later deposits, clays and sandstones, have been laid down in the Vale of York and Mowbray directly over the Millstone grit. Over wide areas the clay itself is of great depth and is full of boulders; it was deposited by slow-moving rivers of ice that carried, even from Shap summit, rocks that are scattered over the plain

The Bridestone, Thornton Dale

far from their place of origin. So many are these boulders, great and small, that in places they are gathered for paving: in Richmond, for instance, the hard cobbles that cover the streets are granite lumps transported long ago down Swaledale from Stainmoor.

This bending of the strata by which the coal measures have been worn away is, as regards money, most unlucky for our county. The accompanying sketch shows roughly what the result has been: a deep cutting from Durham to the West Riding would show that the

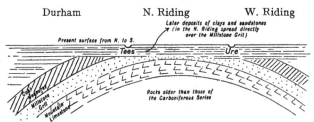

How the North Riding lost its coal

North Riding now occupies an interval in what was once a continuous field of coal.

The thickness of the clay that covers the plain has not been measured. But it must be very great, for in some places people who knew little geology have bored in search of coal, and one such boring on the edge of the Vale of Pickering, at Wass near Ampleforth, was through 400 ft of clay—Boulder and Kimmeridge Clay, a shale belonging to the Upper Oolite. On the coast also where the clay is exposed there is great thickness. For instance, in the artificial cutting on the face of the Castle Hill at

Scarborough it has a depth of 120 ft where it has been scarped for the new marine drive.

In the eastern half of the county are rocks formed more recently than the Carboniferous. These are called in the table Jurassic: of these rocks the North Riding has the greatest area in England. They cover a vast, nearly circular tract including the Cleveland, Hambleton, and Howardian Hills and the Vale of Pickering, and are estimated to be about 3000 ft deep. They are most important for us; for one of the smaller divisions of the Jurassic, the Middle Lias, contains the ironstone worked from Staithes and Skinningrove inland to Eston Nab The Upper Lias, from Robin Hood's Bay to Whitby and inland to Roseberry Topping, contains alum and jet shale, but these are of little import now.

The Lower Lias is seen in fine exposures at Redcar, Saltburn, and Staithes, but especially at Robin Hood's Bay. Here the scars exposed at low water exhibit the rocks of this division nearly to the lowest; and even of this, boulders torn from the sunken reefs are sometimes thrown ashore. The flat-topped heights north of the Vale of Pickering are of oolitic sandstone above the Lias: at Burton Head (1489 ft), whence the Dove runs down Farndale, is the greatest elevation of the oolitic in England. The more recent rocks in the north-east, the New Red Sandstone of the Cleveland area, contain the deposits that have given rise to the salt industry of Cleveland.

We have spoken so far of rocks laid down in level sheets under water, from the detritus of older rocks borne

Names of Systems		Subdivisions	Characters of Rocks
TERTIARY	Recent Pleistocene	Metal Age Deposits Neolithic ,, Palaeolithic ,, Glacial ,,	Superficial Deposits
	Pliocene	Cromer Series Weybourne Crag Chillesford and Norwich Crags Red and Walton Crags Coralline Crag	Sands chiefly
	Miocene	Absent from Britain	
	Eocene	Fluviomarine Beds of Hampshire Bagshot Beds London Clay Oldhaven Beds, Woolwich and Reading Thanet Sands [Groups	Clays and Sands chiefly
SECONDARY	Cretaceous	Chalk Upper Greensand and Gault Lower Greensand Weald Clay Hastings Sands	Chalk at top Sandstones and Clays below
	Jurassic	Purbeck Beds Portland Beds Kimmeridge Clay Corallian Beds Oxford Clay and Kellaways Rock Cornbrash Forest Marble Great Oolite with Stonesfield Slate Inferior Oolite Lias—Upper, Middle, and Lower	Shales, Sandstones and Oolitic Limestones
	Triassic	Rhaetic Keuper Marls Keuper Sandstone Upper Bunter Sandstone Bunter Pebble Beds Lower Bunter Sandstone	Red Sandstones and Marls, Gypsum and Salt
PRIMARY	Permian	Magnesian Limestone and Sandstone Marl Slate Lower Permian Sandstone	Red Sandstones and Magnesian Limestone
	Carboniferous	Coal Measures Millstone Grit Mountain Limestone Basal Carboniferous Rocks	Sandstones, Shales and Coals at top Sandstones in middle Limestone and Shales below
	Devonian	Upper } Middle } Devonian and Old Red Sand- Lower } stone	Red Sandstones, Shales, Slates and Lime- stones
	Silurian	Ludlow Beds Wenlock Beds Llandovery Beds	Sandstones, Shales and Thin Limestones
	Ordovician	Caradoc Beds Llandeilo Beds Arenig Beds	Shales, Slates, Sandstones and Thin Limestones
	Cambrian	Tremadoc Slates Lingula Flags Menevian Beds Harlech Grits and Llanberis Slates	Slates and Sandstones
	Pre-Cambrian	No definite classification yet made	Sandstones, Slates and Volcanic Rocks

down by streams into great estuaries or from the shells of marine creatures dropped miles deep into the sea. There are, however, other rocks formed by the cooling of molten matter that has burst through to the surface from the hot interior of the globe. These volcanic rocks, being due to fire, are called igneous rocks. They usually underlie the more recent rocks; but in places where the latter have been worn away they may form even the summits of the mountains. It is possible that the whole of the North Riding is on the shelf of hard basalt called the Whin Sill. In two small areas of great interest this floor of the county makes itself apparent: in upper Teesdale the basalt has burst through the Mountain limestone, and on the Cleveland moors another crack has given passage to the igneous rock. Along the Tees between Cronkley Fell and Mickle Fell the basalt in its molten state has fired the adjoining limestone and turned it into white crumbling material. The Cleveland Dyke in the east is the longest line of intruding rock traced in Britain. It crosses the river Esk at Grosmont and stretches thence in an almost straight line to the north-west, reappearing beyond the Tees. At Grosmont it is like a great wedge with the edge uppermost, and the sandstone that has been shouldered aside is scorched and partly melted into a glassy rock.

In striking contrast to these ancient volcanic rocks are those that have been formed in our county since the time of glaciers. These are sands and gravel such as we see deposited in the North and South Bays at Scarborough and along the north Cleveland coast.

7. Natural History.

The geologist from his study of the rocks is able to tell us that our county, like the rest of the British Isles, has been alternately depressed and elevated. During long ages it has been below water; for the coralline rocks that rise above the Vale of Pickering were formed from shells in undisturbed water, and the Millstone grit is the compacted dust of older rocks that was deposited in an ancient sea. But also during long ages it has been raised above the sea, for layers of rock like the coal measures once covering the Millstone grit have had time to weather away. Moreover, as many facts go to prove, there was a period—many centuries ago, indeed, but late in the history of the globe—when our country had a climate as cold as that of Greenland. Why this was so is not certain; but we know, from the boulders strewn amid the clay of the plains, from the great scratches on the rocks—high on the shoulders of the Cleveland Hills for example—and from the curious course of the Derwent, that much of the riding was once ice-covered and that slow-moving glaciers ground their way across the land.

During this ice age—the "Glacial Period"—little animal or plant life could exist in our land, but here and there, high up on the hills, are rare "alpine" plants, survivors of the scant vegetation of this cold period. As the ice-fields slowly receded, the land was restocked from the Continent. For the British Isles have not existed as such for any great length of time as measured by geology; between England and the Continent the sea is so shallow that nowhere in Dover Straits could one entirely submerge St Paul's Cathedral. It would take but a small difference in sea-level to make a land connection between Britain and the Continent. Great Britain and Ireland are examples of what geologists call recent continental islands. They were once joined to the mainland, and animals and plants migrated northwards as the climate became warmer. When the sea broke through the isthmus that joined the chalk cliffs of Kent to those of France we had already received, as a result of this migration, most, but not all of the animals and plants found in Europe. As we might anticipate, Britain has fewer species than France or Belgium, and Ireland has fewer than Britain. We should note, however, that the preservation in our land of great tracts for hunting purposes has given the smaller birds safe breeding places that are denied them on the mainland; so that, though we have not so many species, our wild birds are probably far more numerous individually than those of Europe.

Many kinds of animals and plants have in our own county become extinct or greatly lessened in numbers— undesirable kinds like the wolf and hyaena, harmless kinds like the red and roe deer. The beautiful woods of Arncliffe (*erne* = an eagle), through which the Esk runs to Whitby, were doubtless once a resort of this bird of prey and Ravenscar of ravens, but the eagle and the raven are now rare visitors to the North Riding. Exploration of the remarkable Kirkdale Cave showed that at least twenty-two species of wild animal, the tiger, bear, and elephant among them, once lived here. The only wild animals of any size now left in any number are foxes, carefully preserved to provide sport for the many hunting packs. The otter is fairly common along the western streams; but the badger, the pine-marten, and the pole-cat are almost extinct. In the east Tees district conditions are, of course, unfavourable to plant growth; the pollution of the streams and the fumes of the furnaces destroy vegetable as well as fish life.

The North Riding, however, has so extensive and so varied a surface—the mountain moorland with its copious rain, the productive plain, the deeply-carved dales—that most British plants and animals find a home here. Some parts of the county are, indeed, famous hunting grounds for the botanist. One field is in the north-west, from Mickle Fell to Cronkley Fell, where the limestone has been turned by the heat of the igneous rock into loosely-grained (or "sugar") limestone; and here growing in the wiry turf over the white crumbling rock are rare plants like the bright blue spring gentian (*Gentiana verna*), and the evergreen juniper. About the steep basaltic cliffs called Cronkley Scars is found the yellow shrubby cinquefoil (*Potentilla fruticosa*). In the same region on the ridge that encloses Caldron Snout the alpine or yellow-eyed forget-me-not (*Myosotis alpestris*) may be found in July and August, and this seems to be one of the few localities in Great Britain for the flower. It is a survival of the Arctic plants that once grew in the lowland but retreated to

Egg Collecting, Bempton Cliffs

the mountains as the climate grew warm. There is another noteworthy mountain plant in a small glen called the Hole of Horcum, branching off from beautiful Newton Dale, through which the railway from Pickering to Whitby runs. This is *Cornus suecica*, known south of the Highlands of Scotland only here and in the Cheviots. The rare golden saxifrage (*Saxifraga Hirculus*) grows near the joining place of Black Beck with Balder, and where the limestone begins one finds the lady's slipper (*Cypripedium Calceolus*), which does not grow on the grit.

Semmerwater, the curious lake at the foot of Addle-borough, is another productive hunting ground. Here, at the end of summer, specimens may be found of rare flowers like the peculiar lesser bladder-wort (*Utricularia minor*) with its keel-shaped spurs of yellow flowers; and ascending the slopes one may find the crow garlic (*Allium vineale*). In river pastures near Leyburn the autumnal crocus is so plentiful that it makes purple patches on the green. The eastern dales also in many cases delight the botanist. Around Thomassine Foss, a charming fall near Goathland, for instance, the royal fern (*Osmunda regalis*) is still plentiful, though the hand of the collector has been heavy.

There are no more noted grouse-moors than those of Yorkshire, and Stainmoor is famous in this respect. Though not so numerous as they are in the East Riding, the sea birds that frequent the rocky coast north of Scarborough are still a wonderful sight, especially in the breeding season, when the cliffs are haunted by

vast numbers of guillemots, razor-bills, and other
species, whose eggs at one time afforded a regular means
of livelihood to professional cliff-climbers.

8. Along the Coast.

When speaking of the geology of the North Riding
we said that the rock layers from Redcar onwards are
inclined southwards, so that as one travels north from
Filey or south from Redcar different kinds of rocks
appear. These rocks belong to the Jurassic division and
are not hard like granite. They do not, therefore, as do
the Cornish cliffs, break into sharp rugged features;
but still, as they resist unequally to the wearing of water
and weather, they give much variety to the coast.
Some stand firm against the wearing forces—mechanical
forces like the hammering of pebbles driven by the
waves, or chemical forces causing the loss of parts soluble
in water—others waste rapidly. The cliff wall is therefore by no means monotonous, and at places like Runswick Bay, Staithes, and Ravenscar the coast scenery is
hard to equal.

The main thing to note about the North Riding
coast is the mighty wall that confronts the sea from
Redcar to Filey[1]. Its continuity is broken only where
the short becks, or in one case the longer Esk, break
through the ridge by deep wooded glens; and in one
place the cliffs rise higher than elsewhere in England.

[1] The plateau that occupies the east of the county rises abruptly from
the coast.

This is the stretch between Skinningrove and Staithes, which reaches in Boulby Cliff below Loftus the great height of 666 ft. A grander, though not so lofty a mass of rock defends Robin Hood's Bay on the south, where the central ridge of the Cleveland Hills meets the sea. The coast-line from the Bay to Ravenscar climbs steeply up this mass, affording many beautiful views. The railway from Scarborough to Whitby runs close to the edge of the cliff wall; and a traveller is almost constantly in sight of sea, stretches of heather-clad moor, and—since deep water is close in—of ships small and great. But the infinite variety and beauty of the coast can be appreciated only by a more leisurely examination than is afforded by a railway journey.

The rock-bound coast is not hospitable to shipping. The amazing growth of Middlesbrough is partly due to the fact that from the Tees to the Humber there is no port adapted to modern vessels. Scarborough and Whitby, suitable enough for small craft, cannot yet provide facilities for the great steamships of to-day, though in earlier days they were ports when Middlesbrough was a deserted and dismal river flat. Scarborough Castle, standing on a precipice between two bays and accessible from the land only by a narrow neck, was placed where it is in order to command the only harbour in the south; and Whitby towards the north was a famous port in Captain Cook's day, when it sent its home-built ships to the whale-fishing, ships which had a wide reputation for staunchness and sound workmanship.

East Cliff, Whitby

But Whitby, like Scarborough, is handicapped by the cliffs that flank its approach. It was at Saltwick Nab, which projects south of the Abbey rock, that the hospital ship *Rohilla* was wrecked during the early months of the Great War. Runswick and Robin Hood's Bays, beautiful indeed but exposed to the sea and difficult of access from the land, cannot attract the trading ship. The lovely little "wykes," for which the North Riding coast is noted, provide excellent objects for excursions from Scarborough or Whitby but are of slight use to the county. *Wyke* is in fact the Yorkshire form of the Norwegian *vik* or creek, whence the Viking sailed to ravage or to settle, and is usually a deep recess backed by high cliffs on the ledges of which the fishermen's houses find precarious footing.

A voyage along the coast from Middlesbrough to Whitby, 27 miles away, and thence to Scarborough, 17 miles farther, is an interesting one. For the first eight miles to the sea one sees everywhere signs of the successful struggle to make the Tees a ship canal. The huge breakwater, the South Gare, 2½ miles long, is the most conspicuous work of the Tees Commissioners. It was on the "drift" at the Tees Mouth, the broad sandy stretch through which sluggish channels entered the sea, that the men of Cleveland in their northern Ely made a camp of refuge against the Conqueror. The marsh is now reclaimed and great steel-works have risen where the Yorkshire folk made their stand.

The low flat shore continues till we pass Redcar. During the whole way through the Cleveland Hills, the striking form of Eston Nab and the woods on Upleatham rise steeply from the plain; and at Saltburn, where Skelton Beck breaks through a deep wooded gorge, the bold cliff wall begins. The coast here and the coast resorts, Redcar-Coatham, the northern Margate, Marske, and Saltburn itself, are appendages to the industrial region of Cleveland, and have grown with its rapid growth. They are seaside suburbs of Middlesbrough, and are themselves being invaded by "industry." Redcar and Coatham have blast furnaces, and even pretty Skinningrove has its pier for loading pig iron. The mouth of Kilton Beck here marks the seaward end of the ironstone.

Staithes, past Boulby Cliff, is one of the many queer fishing villages along the coast. Its houses are perched like martins' nests on the slopes of the cliff and subsidence into the sea seems to be always imminent. The name of the place is, in fact, that of the sea-walls—the staithes—built against the waves. Runswick Bay, one of the most charming parts of the Yorkshire coast, also has its picturesque village near a wood-sheltered hollow. It is reached from Hinderwell above the cliffs by the steepest of paths.

Whitby and its adjunct four miles north, the little village of Sandsend, are delightful places. Picturesque in itself, with its red roofs and quaint gables, nestling under the cliffs and dominated by the noble ruin of its Abbey on the east, Whitby is as beautiful as any town on the English coast. It stands at the mouth of the Esk, along which by rail and road access is easy to the woods of Arncliffe, and its museum shows a fine series of the fossils found in the neighbourhood.

Robin Hood's Bay, the largest on the coast, stretches between Bay Ness on the north and the Peak or Ravenscar on the south. The cliffs, except on the south, are here low and the villages, of which the chief is Bay Town,

Staithes

do not seem so perilously placed as do Staithes and Runswick. Hayburn Wyke, a favourite resort from Scarborough, is perhaps the prettiest of the Yorkshire creeks.

Scarborough, both the "popular" part along the North Bay and the "fashionable" part along the South Bay, has a perfect situation. The bold Castle rock, now rounded by the marine drive, stands north of the ancient fishing and trading town and the famous Spa with its multiple attractions, and inland lies a stretch of fine moorland.

9. Coastal Gains and Losses. Harbour Works.

Geology teaches us that even the hard Millstone grit is the débris of older rocks laid down under water, and that the mountain masses in the west, half a mile high in parts, were at one time a newly-raised sea bed. We know, too, that much of the North Sea area was once dry land over which animals and men crossed from the Continent. There is no stability either in land or in sea.

After the great convulsions of the earth ceased, slower movements still went on, and they are not even yet ended. In Yorkshire itself there are places actually mentioned in history which have disappeared: Ravenspur, where Henry of Bolingbroke landed, has sunk below the waters. The loss—the coast erosion and coast depression—in the North Riding has been less serious than it has been farther south; though at Staithes, which still clings precariously to the cliffs, the grocery shop in which Cook served as apprentice in 1740 has been washed away, and in 1737 the Spa waters of Scarborough were for a time lost through the subsiding of the cliffs. At Redcar and off the Tees mouth the trees of submerged forests have been found; and in excavations off Middlesbrough peat deposits have been laid bare in which large trees have been embedded. In other parts the firm land has gained on the sea. At Saltburn during diggings for a bridge over the beck a beach raised 30 feet above the sea was uncovered; the shells were all of creatures living since the ice age. On Cat Nab the same bed could be traced.

The work of rivers in shaping the land and filling the sea bed is always going on. There are great flats bordering the Swale and its feeders in the Vale of Mowbray built up by the material brought by the streams from the hills. If we visit the Swale near Ainderby we can see exhibited in miniature what has happened in the past on a larger scale. The river tries to straighten its course; on the inner part of the curve the stream runs swiftly and eats away its bank; on the outer curve the sluggish water, unable to carry its burden farther, deposits sand and gravel.

Man wisely strives to make land and water better fitted for his service; and in the North Riding he does it very successfully. Along the lower reach of the Tees much that was once neither navigable water nor good dry land is either deep waterway or docks or suitable sites for workshops. Some thousands of acres of what eighty years ago were dismal mud flats have been reclaimed, and what was worse than useless is now of value[1]. And the river, banked in by huge containing walls, now runs in

---

[1] This value won from the waters is divided rather curiously ; the Tees Commissioners who do the work obtain half, the Government takes a quarter, and the owner of the land behind takes a quarter.

South Gare Breakwater

a deep channel through the area where once it spread itself in four shifting and tortuous channels. The strange S curve it made below Thornaby has been cut through; and now the Tees, once practically useless for modern shipping, is one of England's great waterways.

The containing walls are built of what would otherwise be waste slag and the river at low water is confined within them. South of the wall reclamation proceeds. The entrance to the river is protected by the magnificent South Gare breakwater, 2½ miles long, built of concrete upon a foundation of slag. At its extremity the lighthouse flashes a red light towards the south-east, whence entry is dangerous, a white light towards the rest of the sea. It may almost be said that the port of Middlesbrough is the creation of the Tees Conservancy Commissioners, and their excellent work is yet in progress.

10. Climate and Rainfall.

The climate of a country or district is, briefly, the average weather of that country or district, and it depends upon various factors, all mutually interacting; upon the latitude, the temperature, the direction and strength of the winds, the rainfall, the character of the soil, and the proximity of the district to the sea.

The differences in the climates of the world depend mainly upon latitude, but a scarcely less important factor is proximity to the sea. Along any great climatic zone there will be found variations in proportion to this proximity, the extremes being "continental" climates in the centres of continents far from the oceans, and "insular" climates in small tracts surrounded by sea. Continental climates show great differences in seasonal temperatures, the winters tending to be abnormally cold and the summers abnormally warm, while the climate of insular tracts is characterised by equableness and also by greater dampness. Great Britain possesses, by reason of its position, a temperate insular climate, and its average annual temperature is much higher than might be expected from its latitude. The prevalent south-westerly winds cause a movement of the surface-waters of the Atlantic towards our shores, and this warm-water current is the chief cause of the mildness of our winters.

Most of our weather comes to us from the Atlantic. It would be impossible here within the limits of a short chapter to discuss fully the causes which affect or control weather changes. It must suffice to say that the conditions are in the main either cyclonic or anticyclonic, which terms may be best explained, perhaps, by comparing the air currents to a stream of water. In a stream a chain of eddies may often be seen fringing the more steadily-moving central water. Regarding the general north-easterly moving air from the Atlantic as such a stream, a chain of eddies may be developed in a belt parallel with its general direction. This belt of eddies, or cyclones as they are termed, tends to shift its position, sometimes passing over our islands, sometimes to the north or south of them, and it is to this shifting that most of our weather changes are due. Cyclonic conditions are associated with a greater or less amount of atmospheric disturbance; anticyclonic with calms.

The prevalent Atlantic winds largely affect our island in another way, namely in its rainfall. The air, heavily laden with moisture from its passage over the ocean, meets with elevated land-tracts directly it reaches our shores—the moorland of Devon and Cornwall, the Welsh mountains, or the fells of Cumberland and Westmorland —and blowing up the rising land-surface, parts with this moisture as rain. To how great an extent this occurs is best seen by reference to the map of the annual rainfall of England on the next page, where it will at once be noticed that the heaviest fall is in the west, and that it decreases with remarkable regularity until the least fall is reached on our eastern shores. These western highlands, therefore, may not inaptly be compared to an umbrella, sheltering the country farther eastward from the rain.

The above causes, then, are those mainly concerned in influencing the weather, but there are other and more local factors which often affect greatly the climate of a place, such, for example, as configuration, position, and soil. The shelter of a range of hills, a southern aspect, a sandy soil, will thus produce conditions that may differ greatly from those of a place—perhaps at no great distance—situated on a wind-swept northern slope with a cold clay soil.

The Meteorological Office in London collects records of temperature, rainfall, direction of the wind, hours of

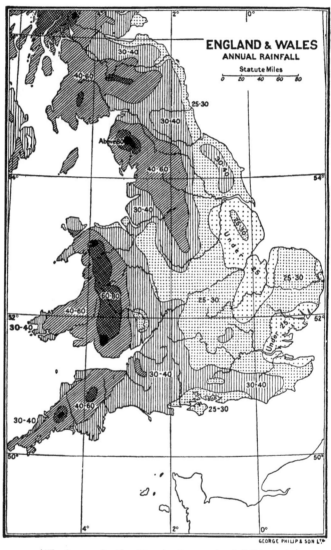

ENGLAND & WALES
ANNUAL RAINFALL

Statute Miles
0   20   40   60   80

30-40

40-60

25-30

30-40

Above 80

40-60

30-40

Under 25

30-40

60-80

40-60

30-40

25-30

Under 25

30-40

30-40

40-60

30-40

30-40

25-30

GEORGE PHILIP & SON LTD

(The figures give the approximate annual rainfall in inches)

sunshine, etc., made by many observers stationed in various parts of the country; and at the end of the year the averages or "means" are worked out for Great Britain as a whole as well as for counties and agricultural divisions.

Let us now turn to our own county, and consider its latitude. The parallel of 54° N. cuts off the little angle of land to the south, and the line of 54° 30′ passes through Whitby in the north. It is thus much nearer to the North Pole than to the Equator. But from the reasons already stated its weather is much warmer than we should expect from its latitude; though, of course, the North Riding is not so favourably placed as the western districts like Cornwall and Wales. Still, we may weigh the importance to us of these warmth-bearing winds by comparing the two great ports, Middlesbrough and Montreal. The Canadian port is about 600 miles nearer the Equator than is Middlesbrough. Yet for five months in the year ships cannot reach Montreal; the St Lawrence is frozen over and navigation ceases. Middlesbrough on the other hand is an ice-free port; and though a very rigorous winter may spread a thin covering of ice over a dock and form a ragged fringe along the shore we need no ice-breakers to keep the port clear.

The average degree of heat at Scarborough, which we take as representing sea-level, is 47½° Fahr. or 15½ degrees above freezing-point. This is called the mean annual temperature, and it of course shows slight variation from year to year. The mean temperature of 1914 was 49½° Fahr.

But almost as important as the high mean annual temperature is the fact that the average temperature of the hottest month, July, is not many degrees above, and the average of the coldest month, January, not many degrees below, the average for the year. For July it is 60°, for January 38°. Thus even in the coldest month the temperature is on the whole above freezing point. We express this by saying that the North Riding has a small range of temperature, or that it has an equable or "insular" climate. If we contrast this insular climate with the "continental" climate of Montreal, we are astonished at the difference. Though Montreal is so much farther south, yet during January it has a mean temperature of 25° Fahr., or seven degrees below freezing, and in July people are sweltering in a mean heat of 70°. The North Riding has not, however, so equable a climate as many other parts of Britain. The Western Isles of Scotland are warmer in winter, though cooler in summer. Milford, which has about the same temperature in summer, is five degrees warmer in winter, and a snowfall is there a rarity.

As the land rises towards the moorlands the temperature decreases; and if we ascend one of the western dales we may note how the crops change from wheat to oats or barley, from barley to pasture, from pasture to bleak heather or grass-covered uplands fit only for the mountain sheep. Hawes, the chief town in upper Wensleydale, though protected by the sheltering walls of the mountains, has a temperature three degrees below that of Scarborough. On the fells themselves the winter cold is intense and we can well understand why the custom began of blowing a forest-horn as a guide to travellers. In Bainbridge at the foot of Addleborough this custom is still maintained during the winter months, and with good reason, for exposure on the uplands through a winter night would very likely mean death.

The Moors above Hutton-le-Hole

The plateau north of Pickering gives another illustration of how position affects climate. The southern slope towards the sun is almost entirely cultivated; the northern slope is a dreary heather-covered upland.

Most of our rain, as we have seen, is brought by the prevailing winds from the west, and as the clouds are forced up the mountain slopes their moisture becomes condensed. The west of the county receives more rain than the east, and the highlands more than the plain. The station at Hawes Junction, where the railway climbs up 1135 feet from Wensleydale into Westmorland, has an average rainfall of 70 inches and this is the rainiest part; at Guisborough on the eastern moors it is 31 inches, at Northallerton in the plain it is only 26 inches.

The rainfall even in the plain is ample for agriculture and pasturage, and the crops grown are limited only by lack of heat. The month in which we have the greatest rainfall—as is the case of the greatest part of our islands —is October. The average fall for this month at Scarborough is 3·16 inches. April is the driest month, yet only 1·34 inches less falls than in October. The rain is thus spread pretty uniformly throughout the year: in our country we can never count on any long succession of rainless days.

The North Riding is not favoured in the amount of bright sunshine it receives. In 1914, a year finer than most, the number of hours registered was 1481. Yet the number of hours during which the sun is above the horizon is 4450. Even when no rain is falling the moisture in the air at times forms mists over valleys and moorlands; and where the furnaces pour out smoke and fumes a darker veil obscures the sun.

Fogs occur during the late autumn and winter. The farmer is fortunate in this respect that the hours of sunshine come mainly during the spring and summer, when his crops are growing and ripening. The fogs are not nearly so frequent as in South Lancashire nor so dense as those of London. The keen, bracing east winds are very trying in April and May, when they blow more often than the west winds; and during the spring months, except in favoured spots, the weather is bleak and cheerless.

It is a matter of importance for the fishermen that the North Riding coast comes within that region of our islands least visited by storms. The two chief storm-tracks pass one up the west coast of Scotland, the other up the English Channel. But calm days are rare, and at times a north-easterly gale renders a voyage along the coast a hazardous one.

11. People—Race, Dialect, Settlements, Population.

Probably no English county has been overrun by more waves of invasion than has the North Riding. Britons, Anglians, Danes, Normans, have in succession ruled. No doubt, the different races that have peopled the county have all left traces of themselves in the looks and character of the present occupiers of the land. But an attempt to decipher these traces would only lead to profitless guesswork. All we can say with certainty is that the Yorkshireman, like other composite races, is a good stock. No doubt, too, the fact of his living where Nature is not very genial, where he has to wrest his means of livelihood from her by force, has helped to make

him the hardy, resolute, and resourceful man he is.
How great a share race has had, how great his sur-
roundings, it would, however, be impossible to decide.

In the case of much of the population, too, notably
in that of Middlesbrough, the men and women are
new-comers from the iron districts of Stafford and
Glamorgan, Newcastle and Glasgow, drawn to the
industrial area by the insistent demand for skilled
workers. Old customs and peculiar dialects are drowned
by the flood of new arrivals. The multiplying of
means of travel and the provision of schools have also
worked towards the diminishing of peculiarities, either
of speech or habits. The broad Yorkshire vowels are
still common and so are many of the words derived from
Danish, for it is possible that the dwellers in this
northern part of the Danelaw were long a Danish-
speaking people. But these are dying out, together with
distinctive customs, through the increase of means of
travelling. The time has gone when at Staithes everyone
was related to everyone else, or at Hawes half the
people were Metcalfs who had to be distinguished by
nicknames.

The typical North Riding countryman—one would
be over-daring if he tried to characterise the many-sided
townsman—is a hard-working, clear-thinking man. He
is keen at a bargain but is generous in his hospitality; he
delights in dogs (his Yorkshire "tyke") and horses; he is
fond of sport, healthy, and long-lived, and is independent
even to aggression. Shrewd and keen-witted as he is,
he has a voluminous folk-lore, and some strange notions

still linger in the lonely farms, perhaps age-old traditions of "the little men" driven out by the Celts.

It is to the Danes that many peculiar words even yet in use are due. To them it is owing that the boy *laiks* rather than *plays*, *loups* more often than *leaps*, will show you the *ganest* way after asking "wheer's tha ganging til?" When the roads are sticky, as often they are in Cleveland, he takes care of his *claes* when walking over the *clarty* ground, lest his mother should *skelp* him, thereby *larning* him to be heedful. The place names, too, prove how thoroughly the Danes had colonised the riding. The distinctive Danish termination for town, "by," occurs no less than 100 times, Whitby being the chief example; the "kirks" are numerous; in the "dales" are "becks" and "fosses"; the "nab," for a prominent feature like that at Eston, is a form of the Danish *næb*, or nose. "Car," the flat, often marshy ground covered with brushwood, is the Old Norse *kiorr*, and appears in Muker, the Carrs along the Derwent, and perhaps in Redcar. Finkle Street in Richmond and in Thirsk is the angle or corner street (Dan. *vinkel*). The war god himself, whose name we perpetuate in Thursday, has given name to Thormanby.

The population of the North Riding in 1911, according to the census taken in that year, was 419,546, this being an increase of 11·2 per cent. on the population of 1901. The growth is due almost solely to the small industrial area in the north-east corner, where half the people are massed. Middlesbrough itself has just over a quarter of the whole and is still growing, though not with its

former startling rapidity. Its estimated population in
1913 was 121,860, including that of North Ormesby,
recently added.

If it were not for this populous district, of which the
seaside places, Redcar and Saltburn, may be regarded
as suburbs, the county would be as sparsely peopled as
Westmorland. Even including the Middlesbrough area
the Riding has only 197 persons to the square mile,
whereas England as a whole has 618. Over its borders
it has Westmorland, the least, and Durham, one of the
most densely peopled of all the counties.

12. Agriculture — Main Cultivations, Woodland, Stock.

Nine-tenths of the great area covered by the North
Riding is occupied by farmed land, though the moorland,
both on the Pennine fells and on the plateau in the east,
is of small value either for the growing of crops or for
the feeding of stock. Where the Millstone grit overlies
the limestone even the black-faced moor sheep fails to
find pasture: the heath-covered stretch is left to the
grouse and golden plover.

But the lowland plain, where the "Drift" deposited
by ancient glaciers lies thick above the New Red Sandstone, is one of the richest grain-growing districts in
England; and leaving the Vale the cornfields ascend
some distance up the fertile dale country. Wheat is
grown freely up to 400 feet (the green-coloured part of

the map inside the front cover) the hardier oats and barley are found up to 1000 feet. As the height increases wheat becomes a precarious crop. The farmer may grow enough for the needs of his family, but on the upland

Farndale

farms the grain will be oats or barley or a mixed crop of both oats and barley (which the farmer grows and grinds together). When specially good prices are being paid for corn, as during the Great War, the ploughed land begins to climb higher up the mountain sides: the number of acres under wheat in 1915 was many thousands greater than in 1913.

Away from the plain, however, the farmer employs most of his land as pasture for his stock during the warmer seasons, or as meadow to provide fodder for the winter. In the dales it is dairy-farming and stock-raising rather than corn-growing; and there is little doubt that the North Riding farmer will in an even greater degree find his profit in his live-stock. We glance down the list of imports into Middlesbrough and among the rest we note a great amount of Danish butter and bacon and an astonishing quantity of condensed milk, Swiss and other. Much of this might well be supplanted by home produce. Already our Wensleydale cheese, made not only in the Ure valley, but also in the districts around, is known to everyone; factors or "badgers" as they are nicknamed, regularly visit the farms to buy butter and poultry, bacon and eggs; and the milk sent by train to the populous West Riding towns, under the encouragement of the railway company, is speedily growing more in quantity.

The North Riding is a grass-growing and grazing county as distinct from an arable county like Cambridge-shire: the two northern counties, Northumberland and Cumberland, alone have more acres of what the Agri-cultural Returns call "Mountain and Heath Land used for Grazing." One acre in every four is thus denoted in the riding; the proportion in Cambridgeshire—a great wheat county—is one acre in 500. Three counties only, Northumberland, Devon, and Kent, possess more sheep than does the North Riding: here they are usually black-faced ("moor" sheep) or white-faced Leicesters. The neighbouring county, the West Riding, has a greater number of cattle than any other county, but the number in the North Riding is not far short. The farmers are eager to improve their stock and the keenest

Yorkshire cattle at Crakehall

interest is taken in the various exhibitions. Few things are more gratifying to the Yorkshire farmer than to win a premium for his shorthorn bull or his Cleveland bay at the All-England show at York. The county takes first rank in the rearing of pigs, of which it has the largest white breed in England.

North Yorkshire, too, is famous for horse-breeding and training. It is whimsically said, "Shake a bridle over a Yorkshireman's grave and he will rise to steal a horse," for he loves horses and is delighted in their exploits. At Middleham on the dry limestone plateau above the Ure are well-known training grounds for

Cleveland Bay Stallion

racehorses; so also are there on the racing flats at Mandale Bottoms on the Tees, at Richmond, and at Malton, "the Newmarket of the North." In the Vale of Pickering valuable carriage horses are bred: the Coach Horse Society stimulates the industry.

Sheep, cattle, and horses showed great increase in 1915 and are likely to reach larger figures still. The increase in the case of corn crops is probably only temporary. Imported wheat is usually so cheap that the English grower cannot sell at prices that pay. He is troubled with a shortage of labour, for the large towns and higher wages attract the country worker. Even before the war the shortage was felt in the North Riding and it is bound to be intensified now. The farmer is therefore impelled to the kind of farming that calls for less labour.

The chief corn crop in the riding is barley, but the area under oats is not much less. The wheat area is little more than a third that of barley. Turnips and swedes are by far the largest root crops, the area for potatoes being only a quarter. Clover and other grasses cover almost as many acres as barley. These latter crops have, as will be noted, a direct bearing upon stock-rearing; turnips constitute a large part of the winter food of sheep and cattle, and a portion of the potato crop is commonly used for pigs, of which nearly every farmer keeps a large stock. It should perhaps be noticed that the acreage of land from which a crop of grass was taken showed a decrease in 1914; but the very dry April and May of that year forced the farmer to devote a greater area of his land than usual to grazing.

Other crops like beans and peas are of slight importance. The small share of bright sunshine is against much fruit-growing and not many acres are utilised for the purpose. Apples are the chief fruit, and even for these the number of acres is less than 500. Devon has over 24,000.

Woodland at one time must have covered a great stretch of the riding, and to-day the lower ends of the dales and the glens on the east are well wooded. Much sound timber is available near Richmond, in the woods of Constable Burton, near Pickering, and in Forge Valley. The North Riding too is the land of great parks. There is, for instance, Duncombe Park of 2345 acres and Castle Howard Park of 1500, and these have many fine trees. But the area of woodland is no more than $\frac{1}{24}$th of the total area: tracts like the former hunting-ground of the kings, the Forest of Galtres, north of York—"Gaultree forest" Shakespeare calls it in *Henry IV*—have been quite cleared of timber.

13.   Industries and Manufactures.

How great a misfortune, financially speaking, the North Riding has suffered by the planing away of its coal measures may be realised by contrasting the population of the county with that of the districts north and south. Durham and the West Riding are two of the most densely-peopled counties; the North Riding is one of the sparsest. The Cleveland industrial area itself is in a manner but an appanage of the Durham coal-field. For the site of Middlesbrough was selected before the working of Cleveland iron-ore had really begun, and the multitude of industries dependent upon cheap iron and steel are still fed by the coal and coke from Durham collieries.

By reason of this lack of easily obtained fuel, what is known as industry, distinct from farming, is almost confined to one small area. But that industry, the smelting of iron-ore and the manufactures due to cheap iron, is a very great one. In the Cleveland area is smelted over a quarter of the whole output of the United Kingdom, that is, about twice the quantity produced in all Scotland. The smelting is carried on mainly at Middlesbrough, but furnaces are in blast also at South-bank, Thornaby, Redcar, and as far south as Skin-ningrove. These blast furnaces with their miniature hills of slag may not be picturesque, but they are signs of strenuous activity in the service of man.

How great that service is we may roughly measure by noting the increase in value as the ore becomes nearer the thing wanted for use. The Cleveland ore at the mine, the greyish green lumps hewn out of the hills at Eston or Boosbeck, is worth 5s. a ton. A ton of pig iron made from about three tons of ore sells at £2. 10s. In the works this becomes transformed into ship plates at £6. 15s., or steel rails at £6. 10s., or boiler plates at £7. 15s. And when, still in Middlesbrough, the steel becomes in one of the three shipbuilding yards part of the structure of a fine vessel, its worth is doubled and trebled.

Smelting is not the primitive process it used to be. At Middlesbrough have been devised improvements in the building of the furnaces and in the treatment of the ore that have greatly cheapened production. Only about one-third of the fuel is now used that was needed in 1851, when the first furnaces were erected. The ore is first calcined or roasted in steel kilns with the result that much moisture and carbonic acid gas is expelled and 100 tons are reduced to about 70. The ore, now dark red in colour, is drawn out into charging barrows.

Blast Furnaces, Middlesbrough

The furnace itself is a remarkable structure between 80 and 100 feet high. At the bottom is the "well" where the molten iron collects. This is built of fire-brick lumps about 3 feet thick and is bound together by stout steel "jacket plates" so that the pressure within may not burst the building. Through the well are pierced holes into which air-pipes ("tuyères") are inserted. Hot air is forced through these in order to help combustion, and cold water constantly circulating round them prevents their melting. Below the tuyères is the "slag notch" for drawing off the waste material that floats on the iron; and at the bottom of the well is the "tapping hole" plugged up with fire-clay till the molten metal is run off into the moulds prepared for it in sand.

Above the well the furnace first widens and then tapers inward to the "throat" which is closed by a steel cone called the "bell." The bell, on which the ore, the coke, and the limestone to form a flux carrying away the waste material, are loaded in due proportion, is raised and lowered by a lever. The furnace is lined with fire-brick and is bound outside with steel straps; for great strength is needed to resist the pressure, the wash of molten materials, and the wear caused by the lumps of ore. And the furnace must, in the interests of economy, run without a stoppage for eight or ten years.

The charging barrows having been hoisted to the "charging gallery" are tipped on the bell, which is lowered so that the materials may slide into the furnace. As the charge descends the ore gradually melts and is at length run into "pigs," though sometimes it is led into huge ladles for direct transport to the steel-works adjoining. There its impurities are expelled by what is called "the basic process," with which the names of Gilchrist and Thomas are associated. Before their discovery pig-iron free from phosphorus was needed for

conversion into steel, and Cleveland iron in which there
is much phosphorus was therefore useless for this purpose.
Now, fine steel is made from the Cleveland ore.

Apart from iron-making and the occupations to
which it is handmaid, manufactures in the North Riding
are of slight importance. There is some linen-weaving
at Yarm and at Northallerton, which has also a linoleum

The Staple Product of Cleveland. Pig iron

factory. A little cotton is spun at Osmotherley and at
Brompton in Allertonshire. Northallerton makes leather
goods and motor cars. Iron in the Middlesbrough area
and farming in the rest of the county are, however, so
far first that the others are negligible.

Some minor industries have vanished. Lead-smelting
has left Swaledale and the alum industry has migrated

to the coalfields. The hand knitters of Hawes could not persist against machine hosiery. Kelp-getting and burning on the Cleveland coast no longer provides a miserable subsistence. Whitby as a centre of the whale fishery was the last stronghold of wooden ships, now superseded by the larger and more economical steel vessels of Tees-side. The port, once famous for vessels of good workmanship like that in which Captain Cook sailed, now merely undertakes repairs; though if the navigation of the Esk were improved a shipbuilding industry might revive. In connection with the whale fishery of Whitby an interesting domestic industry—the making of stays—was carried on at Loftus. Sailcloth and whale-bone purchased from Whitby were made into articles so durable that they were handed down from mother to daughter as heirlooms. This occupation has vanished before the march of modern fashion.

14. Mines and Minerals.

But for the pressure that long ago raised its rocks into a mighty arch the North Riding, as we have seen, would have had rich deposits of coal. As it is, the layers of coal have been weathered away and the county is not well endowed with mineral wealth. Still, the iron ore of the Cleveland Hills is some compensation for the loss of coal, and the salt wells and lime and stone quarries are of value.

Other minerals there are but they cannot be profitably worked. There are still lead mines near Richmond, but the cheapness of the imported lead has caused the once promising industry of lead-smelting to disappear from Swaledale. Many deserted hamlets along the Swale are witnesses of the fact that unless fuel is readily obtainable the reduction of ores will take place on the coast.

Alum mining and burning, too, was of some importance during the seventeenth century, when hundreds of thousands of pounds were sunk in the working of mines along the north-east coast, where the alum occurs in the Upper Lias layer. From Kettleness to Whitby there are still ugly traces of the excavations, and heaps of burnt shale disfigure the cliffs. We are told that alum-mining was begun in Cleveland during Elizabeth's reign. The Lord of Guisborough had noticed in his travels that the vegetation in the region where the Pope had a monopoly of alum mining was of the same vivid green colour as on his estate, and smuggling away miners from Italy he began the industry in Yorkshire. It has now vanished to the coalfields in the search after cheaper fuel.

Jet also is found in the same rocks, the Upper Lias. Its mining must be of immense antiquity, for in some of the barrows on the hills jet ornaments have been found, and Caedmon, the peasant poet of Whitby, refers to the mineral in his poem. The hard variety is perhaps a timber fossil, wood that has been pressed into stone. It occurs in thin layers and is worked with great difficulty. This "real" jet has a fine grain, is very durable, and may be delicately carved and highly polished. Most of that still carved and offered for sale in the Whitby shops is a softer imported material.

The obtaining of salt, another very ancient industry, still employs about 500 men in the Cleveland district. The deposits are here deep down, not near the surface as in Cheshire, and the salt is won from its place under the more recent alluvium by being pumped up as brine. The beds were produced by the evaporation of ancient seas detached by some movement of the earth's crust, and they must be of immense age. Countless ages must have passed after the completion of the salt bed before a thousand feet of rock were deposited over it and, after successive upheavals and subsidences, it reached its present level.

Iron ore, is, however, by far the first mineral. This, too, has been worked long ago, perhaps even by the ancient men who dug the "hut-circles" on the Cleveland moors. It has been mined to any noteworthy extent only since 1850, when a great ironmaster, Mr John Vaughan, found ironstone near Eston Nab. The workable iron-stone is a deposit over about 200 square miles between the Esk and the Tees. It occurs in the Middle Lias, being found in lumps (nodules) and in beds, thin at the coast, but increasing in thickness inland. The bed is cut through by Skelton Beck near Saltburn and a fine exposure made. The mines extend from Skinningrove westward past Loftus, Brotton, Skelton, Boosbeck (where a few years ago a great subsidence took place), to Guisborough. Farther south, high on the moors at Rosedale, a magnetic iron ore is worked: this occurs in the Oolite "dogger" (a local name from the hard stones it contains).

The Cleveland ironstone is not a high grade ore: it takes over three tons to produce one ton of pig-iron. It is therefore much cheaper than the foreign ores brought into Middlesbrough: that from Algeria or from Sweden

Food for the furnaces. Iron ore and limestone

costs about £1 a ton, a ton of the Cleveland ore costs only about 5s. The disposal of the great amount of refuse, or slag, from the ore is a great problem to the ironmaster. Some have tipped the slag on the waste mud banks and so made acres of solid land, some pile up small mountains of it, and some have it taken out into the North Sea and deposited it there. The slag, however, can be turned to use. It may be run into moulds and so make the paving setts for our tram-lines, or may be ground up and mixed with cement to form paving flags, or, again, may be made into "slag wool" which is used for non-conducting linings of refrigerating chambers.

There is an abundance of good stone in the county. In Upper Teesdale the hard volcanic rocks of the Whinstone Sill, and near Goathland and Great Ayton those of the Cleveland Dyke are quarried. The Millstone grit has several good beds of building stone. The stone most quarried, however, is the magnesian limestone. Lime needed for the working of iron ore is obtained near Skinningrove (it is the presence of iron and lime together that has established the iron works in this pretty place), Leybourne, and Pickering. Gannister is quarried on the moor east of Commondale: this is a hard, white, compact sandstone that is unaffected by fire and is much used for the insides of steel furnaces.

15. Fisheries and Fishing Stations.

The sea fisheries of Britain have from very early days been looked upon as a splendid school for seamen, as a kind of foundation for our naval power. Certainly in the Great War our fishing towns have done admirable service. Not only did a multitude of recruits join the Navy, from Redcar, Whitby, Scarborough, and even little Staithes, but also many of the trawlers and their crews entered government service. They left the peaceful gathering of the harvest of the sea and, in the Dardanelles and elsewhere, became mine-sweepers and transports, thus helping the Navy in its difficult task.

For some centuries Britain has been easily first in the fishing industry, and its old predominance is still maintained. The number and size of the vessels, the amount of their catch, and the scope of their operations are steadily growing. Fish figures largely among our exports. In 1913 we sent out over five million pounds worth of cured or salted herrings, Germany buying the largest amount, Russia the next largest. The herring is much the most important of our food fishes, but we also export much cod, pilchards, mackerel, and haddocks. Of most foodstuffs, bread and butter and meat, we produce far less than we need at home; but in peaceful times we sell quantities of fish abroad. From Grimsby, Yarmouth, and Hull fish is carried not only to the Mediterranean and Baltic ports, but to countries as far off as the Argentine Republic. The catch landed at North Riding stations is, however, retained at home; for an excellent railway service carries it speedily to a ready market along the Tees and in the crowded West Riding.

Herring and mackerel, unlike most other fish, are not taken throughout the year. The herring appear in shoals off Wick in May and move slowly southward. In July they begin to arrive off Whitby and Scarborough, and in August and September enormous catches are being landed in the North Riding stations. The Scottish herring fleet visits port after port, followed by large numbers of fish-workers, girls who travel by train to carry out the packing. In October and November Yarmouth is gathering its mighty harvest, and by December the shoals have reached Newhaven. The mackerel, too, appear off the North Riding in the

Fishing Smacks at Scarborough

summer months, and Staithes, which alone of our ports lands any great amount, is then very busy.

After herring, cod is the principal fish landed in the riding and then come haddock, plaice, and ling. Into Scarborough besides, as affording a good market, turbot, soles, and brill are brought. Scarborough is by far the most important of the North Riding stations: then

come in order Whitby, Staithes, Redcar, Robin Hood's
Bay, and Runswick.   But we must not imagine that
any North Riding station has anything approaching the
amount of business of Grimsby or Yarmouth or Hull.
Into Grimsby, the greatest fishing port in the world,
twenty times more fish is brought than into Scarborough;
Yarmouth follows close upon Grimsby, and Hull lands
half as much as Yarmouth.

It is a curious fact, and characteristic of our country,
that our Government spends less money on the fisheries
than any other fishing country—Norway, Holland,
Denmark, Germany, or France.   Yet British fisheries
are more valuable than those of all the rest of Europe
put together.

The east coast of England preponderates greatly
over the west in the fisheries.   One reason is that the
North Sea is cooler than the waters on the west, and the
food-fish are as a rule those living in colder water.
Another reason is that the shallow parts of the ocean
floor, where the fish mostly feed, are more common on
the eastern than on the western shores.   The Dogger
Bank and much of the rest of the North Sea floor is one
great fishing ground, but on the Atlantic side the ocean
rapidly deepens, and fish are both less numerous and
less easy to catch.

This latter advantage, however, is not of much
account nowadays, for half of the fish landed in Britain
are caught in waters beyond the North Sea.   The list
of fishing regions shown in the Government report
includes areas of the ocean stretching from the White

Sea to the coast of Morocco, and these areas are as accessible from Fleetwood as from Scarborough.

The methods of catching fish have of late become so very effective that it seemed likely the supply would begin to fail. Regulations were therefore made, mainly with a view to the protection of the eggs and young of the various kinds, so that more would reach maturity, and these regulations seem to be quite successful in preserving the fish. In the North Riding it is the North Eastern Sea Fisheries Committee that enforces the rules and tries to help the fisherman, and four of its members are nominated by the County Council.

Over three-fourths of the fish landed are caught in trawls, great nets that are dragged along the feeding grounds in the shallow parts of the sea. Herring and mackerel are caught in drift nets, which hang from the surface in fairly deep water across the path of the shoals. Line fishing is nowadays employed only for the larger fish like congers and cod: it accounts for little more than one-fortieth of our total catch. In the larger fishing stations like Scarborough the steam trawler, the finest sea-boat in the world, is quickly superseding the sailing smack.

The inland fisheries are of little importance, though there are some famous angling streams like the Esk above Whitby. But not any are so well stocked as to justify an attempt to make a livelihood by angling.

16.   Shipping and Trade.

Middlesbrough, with its appendage South Bank, transacts nearly all the foreign trade of the North Riding. That is to say, the foreign trade, though great and flourishing, is, like Middlesbrough itself, of quite recent growth. The improvement during the early part of the nineteenth century in means of communication, the building of a railway and the deepening of the Tees, enabled the iron industry to develop in the Cleveland area. For only by rapid and cheap carriage was it possible to assemble the ore, coke, and lime to feed the furnaces. The astonishing growth of industry, attracted by the minerals in or easily accessible to Middlesbrough, is paralleled by the increase of trade.

It was as recently as 1829 that a small syndicate of Quakers bought for £30,000 the site largely composed of bleak salt marshes on which Middlesbrough is built and which is worth very many millions now. One of these "Middlesbrough owners" wrote, when docks for the newly-risen port were first suggested: "We must have them and then we shall bang Newcastle and Sunderland hollow as a shipping place." Middlesbrough has in fact far surpassed both these north-eastern ports in the value of her shipping. It claims indeed to be the best-equipped port in the kingdom; and certainly the wonderful appliances along the quays make for easy, quick, and cheap handling of goods. It is now (1916) the sixth port of England, and as its strange motto

(*Erimus*, "we shall be") shows, looks forward to still higher rank. Like most of the eastern ports it has nevertheless had a severe set back during the Great War: Germany was Middlesbrough's best customer and one of its chief providers.

The port is not on the coast but on what is really a deep ship canal eight miles from the sea. The Tees has by man's skill and energy been transformed from a brawling stream full of rocks and shoals into an important highway. No better example can be found of the active co-operation of man with nature than that afforded by Middlesbrough. It is, in a way, the creation of the Tees Conservancy Commissioners, of whose work we have already spoken.

It is a curious thing that the outward cargoes from Middlesbrough are in most years worth about ten times the inward cargoes. The things sent out are those on which much labour has been expended; the things brought in are mainly raw materials and corn—food for the furnaces and food for the people. The steamer from Belfast, for instance, regularly brings in iron ore from Antrim and takes back steel plates for the Belfast ship-yards. The largest single item of the exports is pig-iron, the moulded bars into which the smelted metal has been run. This we may consider as the staple product of Cleveland; but there are also great quantities of iron and steel goods that have passed through further processes of manufacture—steel rails, bridge-work, machines, steel plates, and so on. A vessel of a famous Japanese line, the Nippon Yusen Kaisha, sails for the East every fortnight laden not only with goods made in the neighbourhood but with general merchandise from Yorkshire and Lancashire—including such things as huge guns from Barrow and Sheffield. One curious fact about this line is worth notice: its newest boats have been built in Japan, but from plates and girders made in Middlesbrough.

Not much coal is exported, though strangely enough it was for coal shipments that the port was established; most of what is loaded into the vessels is for the engines. Salt from both sides of the Tees is a weighty item, as also are the queer by-products, basic slag, used largely in Russia as a manure, and slag wool. A very large fraction—usually about 25 per cent.—of the total shipments are for our Indian Empire. In that vast land Yorkshire rails and locomotives, telegraph wire and bridges, are helping Englishmen to make the country more serviceable to the millions that inhabit it.

Of the imports by far the largest item, more than half of the whole, is iron ore. This is of all kinds, being brought from Spain, Greece, North Africa, Sweden, and even from Nova Scotia. They are all richer in iron than the Cleveland stone; and the pig produced fetches a higher price, for it is more costly to make and is more suitable chemically to certain processes in steelmaking. Corn, flour, and provisions come next; for the farmer of the North Riding has not yet developed his land so that it can feed the people crowded about the works by which they live.

A great many "tramp" steamers frequent the Tees.

These are the common carriers for man on the ocean and have no settled route, going from port to port wherever cargo calls them. Their earnings when plying between places abroad form a large part of our "invisible exports," paid for by sending to Britain food and materials. But there are numerous fine liners, too, including some of the largest cargo steamers afloat, such, for instance, as the British India and the Peninsular and Oriental to India and the East, the Shire to China and Japan, the Union Castle to South Africa, the Milburn to Australia, and the Cairn to Canada.

There was, of course, trade on the North Riding coast and up the Tees long before Middlesbrough was thought of. Whitby not only sent her home-built ships to the whale fishery but used them also for trade; and we often read of landing operations at Scarborough in the bay under the shadow of the castle rock. The chief port, however, in early days was Yarm, now "the only finished town in England," where the building of a house is a startling occurrence. Under the energetic encouragement of the Tees Commissioners and the development of the river as a waterway Yarm may have a bustling future. At present it is a sleepy town, waking only on market days, and its shipping is negligible. It was not always so. There is in existence a list of customs dues paid in 1206, in King John's reign. In it we read that whereas Yarum paid £42. 7s. 10d., Scarborough paid £22, Coatham 16s. 11d., and Whitby no more than 4s. And in an account of the English fleet before Calais in 1346 it is stated that Yarm provided two ships and forty-seven men, Scarborough one ship and nineteen men. The modern port that dwarfs them both was then dismal marshland.

In days when robbers infested the sea, when as in 1377 Scarborough could be victim to a pirates' raid, it was an advantage for a port to be placed far up a stream. But modern commerce calls for depth of water and speed in handling goods, and till these are provided Yarm will continue to languish.

17. History.

We cannot tell in any detail the story of our riding, for to do this would be to rewrite a large part of English history. To an unusual extent men of the North Riding have played a part in the making of England, and many stirring incidents have taken place on its soil. We depend not a little upon conjecture for the history of the riding before it became part of the Roman empire. It is certain, however, that it was inhabited by the warlike Brigantes, "the people of the heights," whose remains are thickly strewn over the north-east moors, and that they were subdued only after severe fighting. In the end, however, the Romans brought the country under civilisation, made roads, built bridges, and organised the people, till then bound loosely together by the tie of common race. Yorkshire became a part of the province Maxima Caesariensis and was garrisoned by the famous Sixth (or Victorious) Legion. The base and arsenal of this Legio Victrix was York, the "other

Rome," the city to which the death of the Emperor Severus in 211 gave a classic celebrity.

The Angles who made descent upon "Cliffland" or entered the county by way of the Humber doubtless found the scattered Britons in North Yorkshire an easy prey. The Romans themselves had sometimes found it difficult to beat off the attacks, and when in 420 the last of the legionaries was withdrawn the land was speedily overrun. Before 550 the English occupation was complete. The North Riding was part of the kingdom of Deira, that pagan land which Pope Gregory, seeing its children sold as slaves in Rome, vowed to rescue from the wrath of God. But it was not till after Deira had been united with Bernicia in 588 that missionaries came. King Edwin of Northumbria, when an outcast, had been befriended by the monk Paulinus and, when king, welcomed his teaching. Paulinus baptised many converts in the Swale and the Ouse, and won Northumbria to Christianity, though after Edwin's death in 633 the land for a little while relapsed into heathendom.

When Oswy, King of Deira and last of the Bretwaldas, defeated the forces of paganism in 655 at Winwidfield, near Leeds, the land was again Christian and twelve monasteries were established. Chief of all was Whitby, under Abbess Hilda, and hence learning spread throughout the land. Here, in 663, was held the synod that settled the fixing of the date of Easter. The missionaries from Lindisfarne taught the ancient Irish usage and this differed from that of Rome. It was at Whitby, too, before the Danes came to devastate it that Caedmon worked.

The Angles from mere plunderers had become con-
querors of Yorkshire. The Danes followed their example,
and during many obscure years incursions gradually
turning into settlements went on. The Old English
Chronicle says little about the north, but we know that
the invaders founded a kingdom in York. This kept
its independence till Athelstane's time. Even when
William the Conqueror landed at Pevensey the land north
of Humber was connected with the south only by the
weakest of ties. During a century and a half the men
of Yorkshire had tolerated in sullen discontent the
overlordship of Wessex. In name a part of England, its
men considered themselves distinct. It was this passion
for independence that brought upon the county so
terrible a punishment at William's hand. Despairing
of bringing to submission men so impatient of rule, the
Conqueror ruthlessly tried to exterminate them.

Harold's defeat at Hastings in 1066 was ushered in
by stirring events in Yorkshire. Tostig, King Harold's
banished brother, had obtained help in his effort to
gain back his earldom from the Norwegian king, Harold
Hardrada. Their fleet "made the coast of Kliflond,
where all men fled"; Scarborough was taken and burnt;
the invaders sailed up the Ouse, and at Fulford, two miles
from York, defeated the forces of the northern earls,
Edwin and Morcar. English Harold, whose help had
been invoked by these half-independent earls, hastened
north, surprised the enemy encamped on both sides of
the Derwent at Stamford Bridge, and attacking from
the right bank won the day.

For some time after Hastings Yorkshire was little affected by the Conquest, and its people supposed that they were to continue as independent as under the previous rulers at Westminster. But William would be content only with a united England. The savage attack on the Norman garrison of York in 1068 gave him the pretext he wanted for enforcing his will. Submission now brought no mercy; with pitiless severity he harried Yorkshire so that no human being might live there. Homes, crops, cattle were all destroyed; the fertile Vale of York became a deserted wilderness and the bodies of those who had perished from famine lay rotting by the roadsides. Long after that terrible winter of 1069 the land bore abundant traces of the ravage. In the Domesday Survey of 1086 entries about the "Nort Treding" reiterate for page after page with ominous brevity "waste" (*vastum est*). The North Riding as in any way under separate rule ceased to exist. Yorkshire and its story became henceforth in reality a part of England and English history.

It was as an Englishman, not as a northern earl, that Peter de Brus, lord of Skelton Castle and ancestor of the "patriot king," helped to force from the reluctant John the grant of Magna Carta in 1215. During King John's subsequent short time of power he besieged and captured Skelton Castle, indicating that Brus's part in obtaining the charter was not small. It was as part of England, too, that until the date of Flodden (1513), the North Riding was an advanced post against Scottish attacks, and as such was never long exempt from disturbance.

During the long and wasteful struggle between North
and South small marauding bands repeatedly penetrated
into Yorkshire, where the fortified church towers still
remind us of the precautions that had to be taken.

In 1138 King David of Scotland invaded Yorkshire
with a horde of wild Highlanders in the cause of his
niece Matilda. The invaders were met three miles from
Northallerton by a force summoned by Archbishop
Thurstan's pastoral letter. Bruce, Baliol, and Mowbray
were there with their retainers, and the steel-clad
knights standing before their archers easily repelled the
Highland attacks. In two hours King David's forces
were routed and he himself in flight. Behind the
English archers, in token that the Church was fighting
against the aggression of savage robbers, was carried a
mast bearing four sacred banners—those of St Peter of
York, St Wilfrid of Ripon, St Cuthbert of Durham, and
St John of Beverley. This it was that gave this fight
the name of the Battle of the Standard.

During Edward II's unlucky reign the riding had
a troublous time. It was at Scarborough in 1312 that
Edward was obliged to desert his favourite Gaveston;
and after Bannockburn the Scots, bold in their success,
paid three visits in force to the county. In 1318 under
Black Douglas, the staunch friend of Bruce, they ravaged
up to Scarborough, which they burnt. In 1319 Bruce,
seeking to divert Edward from Berwick, which he was
besieging, sent forces towards York. At Myton-on-
Swale these defeated an English army hurriedly gathered
by the Archbishop of York and his clergy. So great

was the number of clergymen present at the battle that it was called "The Chapter of Myton." At the battle of Byland in 1322 Edward himself, striving to retrieve the great blow of Bannockburn, was completely routed.

The dismal times did not end with Edward's death, for in 1349 the Great Plague ravaged the land, with

Scarborough Castle

famine in its train. On the road to Bowes from Cotherstone there is still shown the Butterstone where, to avoid infection, money and provisions were exchanged. The changes in religion during Tudor times were not welcome in the North Riding, and in 1536 a rising against Henry VIII's government took place. It was headed by Robert (or Roger) Aske of Aske Hall, near Richmond.

Headed by a procession of priests with crosses and banners—whence the rising was known as the Pilgrimage of Grace—Aske went to besiege Scarborough Castle. But it was stoutly defended and with the approach of Henry's troops the insurgents melted away. Aske and many of his followers were put to death. Among them was the last Abbot of Jervaulx, who was hanged at Tyburn in 1537. Visitors to the Tower of London may still find the name he carved while awaiting execution in Beauchamp Tower, "Adam Sedbar, Abbas Jorevall 1537."

During the Commonwealth wars no battle was fought in the North Riding, though Marston Moor (1644), a few miles beyond the border, was the scene of the great triumph of Cromwell's Ironsides. Prince Rupert, during the night before the battle, had encamped on the edge of Galtres Forest. The men of the riding as a rule were adherents of the King, and the castles of Bolton, Helmsley, and Scarborough were garrisoned by the Royalists. The last castle twice sustained a siege during the contest.

The later peaceful progress of the county—its share in the introduction of railways, its industrial triumph in the creation of Middlesbrough with its multiple activities, its growth in trade, its farming achievements —is detailed in other sections. The Great War, of which one incident was the bombarding of Scarborough and Whitby, has no doubt sadly interrupted its orderly development. But with peace we may anticipate that the North Riding men and women, splendid in war, will speedily retrieve what strife destroyed.

18. Antiquities.

From the scanty relics of the earliest inhabitants of our island, from the things they made and used, the historian tries to build up the story of their lives, just as the geologist from the fossils found in the rocks tells the story of the land. In some of the English counties relics have been found of a very early race of men who hunted in the forests of the Thames Valley and the southern parts of our land during the far-off days when Yorkshire was yet hidden by its ice covering. But probably no remains of the men of the Old Stone Age (or Palaeolithic Period as it is called) have been found in the North Riding.

When after a great gap of time our climate became more genial, the men of the New Stone (or Neolithic) Age appeared. They came, it is supposed, from the shores of the Mediterranean; and so far as we know they were the first men to live in the North Riding. Of their implements, their weapons and tools, and their other possessions, the pottery they made and the ornaments in which they delighted, a quite remarkable harvest has been gathered into our museums. They lived on the higher land on the north-eastern moors, not on the marshy plain or on the loftier fells to the west.

These men must have learnt the use of the bow, for they made finely chipped and polished arrow-heads, leaf-shaped flints with sharp points; they utilised more effectively their ability to throw things by means of

javelins or spears, the heads of which were polished
flint; they flayed and cut up the animals they slew with
long straight-edged flint knives. They had flint scrapers,
too, with which they prepared the skins for wear. To
obtain the material for their weapons they dug deep to
the beds of flint. The women were beginning to spin
and weave; they shaped clay and burnt it into pottery;
and they were fond of ornaments—the first wearers
of jet beads. They introduced the breeds of cattle
prevalent in our country until the Angles came, for the
New Stone men were herdsmen, not merely hunters,
though they knew nothing of metal-working, the art
that has done most for the welfare of mankind. They
built great egg-shaped barrows or funeral mounds, for
they honoured the dead and placed by their side objects
that might be of use in some future life.

The Stone Age men were conquered by a branch of
the great Celtic family that had learnt how to make
hard bronze weapons. Somehow it was discovered
that a little tin mixed with the already known copper
produced a harder metal, and so bronze came into
being. Only very scanty remains of the Bronze Age,
however, have been found in the North Riding. The
tools or weapons made of the metal were so valuable that
they were not buried in the barrow, but were handed
down as heirlooms and carried away before the next
wave of invaders.

These were the Brythons, to whom the name Britain
is due. They were akin to the Bronze Age men, but they
had learnt how to make iron, which is even more effective

in war than bronze, and they drove away or subdued the former owners of the land. We live to-day in the Iron Age with which real history begins. The barrows or howes, entrenchments, and other remains of this age are exceedingly numerous in the North Riding. On the north-eastern moors their vast burial mounds, some of them hillocks, are thickly scattered, and hundreds have been carefully opened and explored by patient examiners. Some on hills overlooking the sea recall Beowulf's dying request to be buried on a promontory that sailors might see his barrow as they sailed afar over the darkness of the floods. On Danby Moor is a so-called British village, a favourite object of visitors from Whitby. A large number of pits, about 12 feet in diameter and as many deep, are yet visible in the main ironstone seam and some have supposed them to be hut-circles. There is no doubt, however, that they are merely early excavations for ironstone. On the moor above Hayburn Wyke is a rough stone circle.

Some remarkable lines of earthworks are supposed to bear silent witness to the progress of this Brythonic invasion. The best known is on Eston Nab, and there are several stretching miles in length, crossing the ridges south of the Esk—at Glaisdale and Westerdale for instance—that are evidently built against some threatened attack from the south. Across Castleton Ridge the "Highstone Dike" exhibits one section faced with stone towards the south and with a deep ditch before it. The purpose of these entrenchments appears to have been the enabling of the invaders to make secure what land they had already won and to give them a fortified base from which to make further conquests. A line of entrenchments known as Scot's Dike stretches from the Swale to the Tees, and another group on the Howardian Hills overlooks the Vale of Pickering.

No English county, except indeed Northumberland, has more Roman remains than has the North Riding.

Wade's Causeway; a Roman Road on Wheeldale Moor

Nearly every part of our county, on the plain at North-allerton, among the western fells at Bainbridge and Bowes, and on the eastern moors at Cawthorne camp, contains traces of the occupation of this wonderful people. The Emperor Hadrian in 120 brought the Sixth Legion from Germany and stationed it at Eboracum. Till the close of Roman rule in our land the

North Riding was full of their activity. The greater
part of the officers and men were engaged in garrison
duty along the northern walls. York was their base
and headquarters, and there was constant traffic from
that centre along the roads. When danger threatened
from the north the *vigiles* (the firemen who kept watch)
at Greta Bridge would fire their signal beacon. Past
Scotch Corner and Catterick, past Northallerton and the

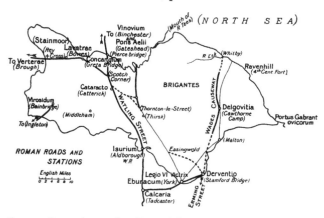

"street" stations, the line of beacons summoned from
York forces to repel the attack.

The sketch map here given shows the site in our
county of the chief Roman stations so far as we can
identify them. They were connected by paved military
roads[1], some parts of which, showing how well the Roman

[1] The road from Aldborough to Piercebridge through Catterick was part
of famous Watling Street ; that from Malton to Whitby probably part of
Erming Street. (See Codrington's *Roman Roads in Britain*, 1918 Edition.)

engineers did their work, still remain after so great a
lapse of time. The thicker lines indicate Roman roads
actually in use now. The chief is that across the county
from Aldborough in the West Riding through Catterick
to Piercebridge on the Durham side of the Tees. At
Scotch Corner the road branched off to cross the dreary
height of Stainmoor, past Bowes (whence a stone
inscription came, now preserved in Trinity College,
Cambridge), on to Carlisle and the western end of the
Wall. The motor riders who to-day tear along this part
of the "Great North Road" are indebted for its straight-
ness to the Roman builders so long ago who, sparing no
time or labour, hewed their way through forests, and
built causeways over the marshes, in order that their
armies might move with speed. As interesting as any
is the road over the fells to Lancaster from the pretty
village Bainbridge, where the Roman camp is still
preserved, unopened and certain to contain abundant
remains. Grass-grown and silent now, the skilfully-laid
stones are there much as the Romans left them, and the
road, scorning to deviate, climbs in a direct line across
the mountain ridge. In York Museum is a magnificent
collection of Roman remains, altars, inscriptions, coins,
and weapons, and in the Dorman Memorial Museum at
Middlesbrough are most interesting relics of the Roman
occupation of Cleveland. These relics—coins, pottery,
ornaments, cloth, etc.—came from the Saltburn Camp,
an outlook fortress at the summit of Huntcliffe, excavated
and explored in 1911–12.

We might expect the Saxon period to yield more

remains of antiquarian interest than the earlier Roman; but with the exception of grave-covers or "hog-backs"— perhaps the finest collection in England is at Brompton-in-Allertonshire—and crosses the work of the Anglo-Saxons has disappeared. All the stone work that can with certainty be described as Saxon is in churches. We must remember that our early ancestors trusted little to walls and disliked confinement in buildings. Their settlements were strong in forest and marsh obstacles rather than in fortification. One such fortified earthwork should indeed be mentioned—that of "William's Hill" beside Middleham Castle. Speight in his *Romantic Richmond-shire* describes it as "the most perfectly preserved example of an Anglo-Saxon *burh* to be found in York-shire." But between Roman and Norman times there was little military building.

19. Architecture—(a) Ecclesiastical.

Before we consider the churches and other important buildings of our county we may say a few words about the various styles of English architecture.

Pre-Norman or—as it is usually though with no cer-tainty called—Saxon building in England was the work of early craftsmen, who had an imperfect knowledge of stone construction. They commonly made rough rubble walls with no buttresses, small semi-circular or triangular arches, and square towers with what is termed "long-and-short work"—the stones being placed alternately horizontal and perpendicular—at the quoins or corners.

Saxon building survives almost solely in portions of small churches like those at Kirkdale, Kirk Hammerton and Hornby. The Saxon tower of Appleton-le-Street church is the finest in Yorkshire.

The Norman conquest started a widespread building

Kirk Hammerton Church

(Showing Saxon portion in foreground and modern portion behind)

of massive churches and castles in the continental style called Romanesque. The style is in England known as "Norman." It is marked by walls of great thickness, semi-circular vaults, round-headed doors and windows, and massive square towers.

From 1150 to 1200 the building became lighter, and the arches pointed. Then was perfected the science of vaulting by which the weight is brought upon piers and buttresses. This method of building, the Gothic, originated from the wish to cover the widest and loftiest areas with the greatest economy of stone. The first English Gothic, called "Early English," from about 1180 to 1250, is characterised by slender piers (commonly of marble), lofty, pointed vaults, and long, narrow, lancet-headed windows. The splendid ruins of Whitby and Rievaulx Abbeys are Early English, but the most perfect example is the little church of Skelton, north-west of York. After 1250 the windows became broader, were divided up and ornamented by tracery, and in the vaults the ribs were multiplied. The greatest elegance of English Gothic was reached from 1260 to 1290, when English sculpture was at its best, and art in painting, in coloured-glass making, and in general craftsmanship was at its zenith.

After 1300 the structure of stone buildings began to be overlaid with ornament, the window-tracery and vault-ribs were of intricate patterns, the pinnacles and spires loaded with crocket and ornament. This style is known as Decorated, and lasted till the Black Death, which after 1348 stopped all building for a time. The finest example of Decorated in the county must have been Guisborough Priory, of which the magnificent east end still overlooks the red-tiled town.

With the changed conditions of life the type of building changed. The style called Perpendicular, unknown abroad, developed with curious uniformity and quickness in all parts of England and lasted from the latter part of the fourteenth century with scarcely any change up to 1520. As its name implies, it is marked by the perpendicular arrangement of the tracery and panels on walls and in windows, and it is also distinguished by the flattened arches and the square arrangement of the mouldings over them, by the elaborate vault traceries (especially fan vaulting), and by the use of flat roofs and towers without spires. Thirsk parish church is perhaps our best example.

The medieval styles in England ended with the dissolution of the monasteries (1530–1540), for the Reformation checked church-building. There succeeded the building of manor-houses, in which the style called Tudor arose—characterised by flat-headed windows, level ceilings, and panelled rooms. The ornaments of classic style were introduced under the influence of Renaissance sculpture and distinguish the Jacobean style, so called after James I. About this time the professional architect appeared. Till then building had been entirely in the hands of the builder and the craftsman.

Of these later styles, except the Decorated and Perpendicular, which flourished while the contentions with Scotland brought much harm on Yorkshire, the North Riding has many and notable examples. The century after the Conquest was dignified and elevated by a wave of religious enthusiasm. Nowhere was it stronger than in the region that William had cruelly ravaged, and abbeys and churches bear eloquent witness of it. Almost every dale together with its castle—mouldering and deserted like Middleham, or still the scene of military activity like Richmond—has its ruined abbey and its ancient church. The Norman style especially is evident. This, as we have seen, prevailed from the Conquest until about 1200; and when we consider the aggregate of the abbeys, churches, and castles erected in England during that short period we must marvel. The total far exceeds the mass of public buildings built during any like period in any country whatever.

Many of the buildings—magnificent abbeys of which the ruins still astonish and delight—were the homes of the monks. One result of the great religious revival of the twelfth century had been the institution of various orders of men who sought holiness by turning from the distractions of the world and living a life of self-denial and submission to rule. Long before, St Benedict had framed rules by following which his monks might lead good, pure lives. It was his followers, the Benedictines, who rebuilt the Abbey of Whitby, investing with new grandeur the site that in Abbess Hilda's day had been the beacon of light and learning in northern England. Placed 200 feet high on a bluff promontory and conspicuous from a wide stretch of sea as well as from miles of heath-clad moor, the glorious beauty of the abbey has appealed to men through the centuries. Other abbeys are in quiet, secluded places and in their retirement are a symbol of the monk's vow of renunciation; but Whitby challenges notice. Its great central tower fell in 1833, succumbing to the sea winds against which it had so long stood unmoved, and the building was still further injured by the German bombardment. There are still left the beautiful north transept, the most magnificent part of the church, the Early English chancel with its seven stately bays, and portions of the nave and north aisle. The Abbey cross is erected near.

Some of the Benedictine monasteries had become places of wealth and ease, and consequently men who wished to carry out more strictly Benedict's rule formed new orders. The chief of these was the Cistercian order, so called because their rule originated at the little monastery of Citeaux in Burgundy. The work of the Cistercians is the most striking feature of the religious life of Yorkshire. In the North Riding they built the splendid houses of Rievaulx, Byland, and Jervaulx. Their rule was strict. Silence had to be kept; they had to labour every day, to rise at two in the morning for the first of the daily services, and to eat only the plainest vegetable food. They wore a dress of undyed wool and so were named White Monks.

They chose some sheltered valley for their settlements. Rievaulx, the mother abbey from which others like the famous Melrose are offshoots, is so beautiful in its ruin that not Tintern itself is more lovely. As in all the Cistercian abbeys the main buildings were grouped around the four sides of the cloister garth. On the north was the church, on the south the refectory or dining hall, on the east the dormitory for the monks, and on the west the dwellings of the lay brothers. Outside the main building were the infirmary and the strangers' chapel—for no stranger was admitted into the abbey. The abbot, who besides ruling the monks was a great noble having lands and tenants, usually had a house outside the abbey.

Rievaulx (Rye-vales) in the quiet valley of the Rye, three miles from Helmsley, is so sheltered in the dale

Rievaulx Abbey

that the wonderful carving of its pleasantly-coloured sandstone is still almost as sharp as when the chisel of the monkish craftsman left it. The church has, very unusually, been built north and south, for the nature of the site has prevented an east and west position. The parts of the building remaining intact are the Early English chancel with its seven bays stretching

to the fine eastern lancets and the transepts. The
daughter abbey, Byland, reached by a moorland
track, three miles over a ridge into an even more
secluded hollow, is of less interest; but the ruined
nave and the great west front with fine Early English

Guisborough Priory

work tell of former stately glories. Jervaulx, near the
foot of Wensleydale, was another Cistercian establishment. Situated on level meadow land south of the Ure
it suffers by comparison with Rievaulx, though its
ruined chancel gives evidence of a fine Early English
building.

At Guisborough, presiding over the quaint market town, is the noble Decorated east face of the once grand Priory. Great churchmen thought that there should not only be places of retirement, but also places which were centres of activity, and they founded orders of clerics who took the vows of an austere life but who lived and worked among the people. They were called Friars or Canons. The Priory of Guisborough, founded by one of the great Bruce family, belonged to the Augustinian Canons. Mount Grace Priory, finely placed in the oak wood of Arncliffe, six miles north-east of Northallerton, is the one house in the North Riding of the more austere brotherhood, the Carthusians. There are indeed only nine more in England, and that in our county is the most complete. The Charterhouse at Mount Grace was, as always, built to ensure silence and separation; the Carthusians ate, worked, and slept apart from their fellows. Each separate cell of the priory has, for instance, a small opening with a double bend, or elbow, in the middle so that food might be passed through without converse.

St Francis of Assisi, preaching a gospel of love, instituted an order vowed to poverty, giving all they had, and in their barefooted wanderings living on alms. These were the Franciscans or Grey Friars. Near Richmond are a Norman doorway and a later tower— a handsome specimen of Perpendicular work—that mark the site of a house of this order.

Easby Abbey on a beautiful stretch of the Swale near Richmond, Coverham, Scarborough, Rosedale,

The Norman Crypt, Lastingham Church

Eggleston made famous by Scott, though all worthy of close examination, can only be mentioned.

The ancient churches of the North Riding attract one less than do its ruined abbeys and castles, but there are some general features that must be noticed, and many buildings are of exceptional interest. Such is Lastingham, secluded on the moors, once a noted Saxon shrine, placed as Bede tells us "among craggy and inaccessible mountains where wild beasts were to be met rather than friendly men." The crypt of the later church is the finest Norman work in the county. In Northallerton church, a grand old structure perhaps due to the prince-bishops of Durham, we have Norman and later Early English styles; some Saxon fragments are treasured and there is beautiful Perpendicular woodwork. Thirsk parish church, a magnificent Perpendicular building, is the finest in the county. Its curious crypt under the chancel is probably owing to the slope of the ground towards Cod Beck, and is not, like that at Lastingham, an inner sanctuary. The church of Patrick Brompton, near Jervaulx, is mainly Early English, but the chancel is an example of the best Decorated work. At Kirkdale, over the south door of the old Saxon church, is a famous sun-dial, 7 feet long and bearing the longest inscription extant of the Anglo-Saxon period. The grand old church of Pickering is famous for a remarkable series of wall-paintings above the nave. These frescoes, dating from the fifteenth century, are the finest in England.

One striking point about the older churches is the

strength of their fortress-like towers. Many, doubtless, have served as places of refuge during harassing raids

Frescoes in Pickering Church

from the north. Some otherwise simple church towers —Bolton-on-Swale and Danby Wiske are instances—

have their lowest stage vaulted, probably to lessen the danger of fire. The rectory itself at Danby Wiske is moated. The doorway to the lower stair at Bedale was defended by a portcullis. The massive structure of the

Spennithorne Church

twelfth century tower at Melsonby is evidently easily convertible into a stronghold; and at Spennithorne the battlements of the tower have borrowed an ornament from military architecture and are crowned by figures of "defenders."

20.   Architecture—(b) Military.

Dwelling amidst what was long a hostile populace
the Normans developed the building of castles into a
science; Richmond and Scarborough thwarted the
bravest efforts of insurgent Englishmen.   And the fact
that the North Riding during many centuries had to
bear the brunt of Scottish raids brought into clearer
light the utility of a fortress that might become a haven
of safety or a base for offence.   Against the frequent
inroads of marauders from the north the farmsteads
themselves were built in the form of a "pele" or
"peel" tower, into the outer yard of which—the
"barnekin"—the cattle might be driven when danger
threatened.   We can thus understand why the riding
has scattered over its surface more ruined strongholds
than any other county except Northumberland.

The great nobles of the North, remote from the central
authority in London, ruled from Richmond or Middleham
or Helmsley as petty kings.   From every point of vantage
their fortified homes dominated the district around:
there were Skelton and Kilton on the northern flanks
of the Cleveland Hills; Mulgrave and royal Scarborough
on the coast—the one on a narrow isthmus between two
deep ravines, the other on a precipitous cliff; the squat
tower of Bowes guarded the pass over Stainmoor, the
one depression between the Tyne and Aire Gaps; and
there was a striking series that kept watch over the dales
opening from the north upon the Vale of Pickering

—Ayton, Kirby Moorside, and Helmsley. But border warfare has long ceased and the domestic is now more pronounced than the military element in a noble's house, and the ancient strongholds have been deserted for more comfortable homes elsewhere. Many of the castles—Axholme, Kirkby Malzeard, Thirsk—have been destroyed in punishment for the rebellion of their lords; others like Mulgrave retain only the name of the ruins near them; of some, like Bishop Pudsey's at North-allerton, not a trace of stonework remains.

The earliest fortresses built by the Normans were lofty stone keeps like that of Richmond. The strength of these keeps depended on the thickness of their walls and the advantages of their position: Richmond is placed on a mound that rises high above the Swale, which almost encircles it, Scarborough on its great Castle Rock is connected with the mainland only by a narrow and easily defended neck of land. The keeps were usually three stories high. The lowest contained a well, often of great depth, and the store rooms; the middle provided the soldiers' quarters; and the third was the home of the governor and his family. The upper stories at Richmond, where the keep is still almost perfect, were reached by straight staircases in the thickness of the walls. Later, one, and afterwards two, outer walls enclosed a space around the keep forming an inner bailey and an outer bailey. If the defenders were driven from the outer wall the inner was a second line of defence before the keep could be attacked. A moat, crossed by a drawbridge, usually surrounded the castle, and the entrance gate was protected by a portcullis that could be dropped at will. The outer defences often enclosed a great space; that of Scarborough is about nineteen acres, sufficient for a small army's camp.

Richmond Castle, "standing fair upon the hill," was built by the Conqueror's nephew to overawe Richmondshire—the princely gift of Swaledale and the Vale of Mowbray. The Teesdale Gap was the usual passage from Scotland, and William, realising the value of having a permanent military force here, authorised the castle. This was most likely in 1069, after the fierce harrying of Yorkshire and Durham. The massive keep, under the shadow of which the town grew, was supplemented by two outer wards.

Scarborough was probably one of the unauthorised —"adulterine"—castles that arose in Stephen's reign. But Henry II compelled its surrender, and it has since been a royal castle, witnessing many stirring sights— among others the surrender of Gaveston in 1312, and sieges successful and unsuccessful during the Pilgrimage of Grace (1536) and in the Civil Wars.

Fortifications at Middleham were also authorised by the Conqueror, though the present castle was not begun until 1190. It is best known as the castle of "the Last of the Barons," the King-maker Warwick, greatest of the Nevilles. Richard III by his marriage with Anne Neville obtained the castle and often lived there, and here was born and died his only son. Middleham is peculiar in having only one narrow space between the

Richmond Castle

keep and the outer wall.   The decay of its rubble stone work is in great contrast to the endurance of Roman building.

Bolton Castle, on rising ground farther up the river and on the opposite side, was built by the Scrope family

The King-Maker's Castle (Middleham)

("Lord Scrope of Masham" is one of the conspirators in Shakespeare's *Henry V*).   It is well known as one of the prisons of Mary Queen of Scots, and there is a tradition that she attempted escape but was recaptured at "Queen's Gap" on Leyburn Shawl.   Bolton is a large example of a rectangular keep (1379) with an open courtyard.

Pickering is an instance of the transforming of an earth and timber castle into a medieval one. The

Bolton Castle

defensible site is on a cliff above the stream two miles from the Roman road from York. The earthworks, entirely of Norman type, were supplemented by the stone building.

21. Architecture—(c) Domestic.

The comparative rarity of notable dwelling-houses in the North Riding forms a great contrast to the wealth of churches, abbeys, and castles in the county. When there ceased to be need of a stronghold to defend his

Rafter-built house showing "crucks" and framework

land the noble no doubt often built his manor-house near the Court and the capital.

A local peculiarity are the curious rafter-built houses on the Cleveland moors, in which the rough stone walls were reared about two wooden forks whose ends rested on the ground. They are nowadays supplanted by more healthful and more spacious dwellings.

The ordinary houses of a town are generally built of materials from the neighbourhood. The cost of carrying goods is small enough nowadays, but heavy materials like granite and marble would be brought from a distance only for special buildings. The wood or stone near at hand must suffice for the houses of ordinary folk, and if none is available bricks from clay are made. So we find in the Millstone grit region towns built of the rough and rather porous stone. Through this district one meets with solid stone-built farm-houses, some as old as Elizabeth's reign; the roofs themselves are of slabs of finer grained and easily split gritstone. Here the very fields are divided by walls of stone and the paths (or "causeys") through them are all flagged. The limestone too is extensively quarried, and thus through the great dales the houses are of stone—are all "little grey homes in the west." At Bedale, however, we reach the brick-built towns. Where the eastern hills provide material, as they do for the long straggling street of Ampleforth, the more picturesque town reappears. In the Cleveland industrial region there is certainly enough clay for brick-making—"Cleveland in the clay, Bring two soles, bear one away"—and red brick is the common building material. The growth of Middlesbrough, for instance, was so rapid that there was time for the erection of nothing but hurriedly-built brick houses, and the eagerly-flocking workers lived in miles of mean streets. But of late a wonderful improvement has taken place in the housing of Middlesbrough people. One reason for this improvement is the wisdom of many workmen who buy a good house through a building society and add to its comfort by their own handicraft.

Here and there in the riding, often as the successors of strong but inconvenient castles, are the seats of great families. Snape Castle near Bedale, long a stronghold of the Nevilles though now only a farm-house, is a fine

Snape Castle

old quadrangle improved by the Cecils of Elizabeth; and is a residence rather than a fort. The modern Mulgrave Castle, a few miles west of Whitby, has no pretensions to strength, but is a magnificent dwelling-house. So, too, is Duncombe Hall, dwarfing the poor ruins of Helmsley Castle beside it. The Hall, rebuilt after being burnt, was planned by the poet-architect, Sir John Vanbrugh, in Queen Anne's reign, and is one of the massive structures that gave rise to the epitaph on him,

> Lie heavy on him, Earth; for he
> Laid many a heavy load on thee.

Castle Howard, a famous show-place four miles west of Malton, also was planned by him. The imposing mansion is one of the finest examples of Corinthian

Castle Howard

Renaissance, and its picture gallery is one of the art treasuries of England.

Other notable country houses are the Elizabethan Brough Hall near Catterick Bridge; Marske Hall, a fine manor-house built near Redcar by the Pennyman family in James I's reign; Burton Hall, the home of the Wyvilles, in a pleasant park at the foot of Wensleydale; Nappa Hall near Aysgarth, the ancient home of the numerous Metcalf family and one of the many

places where the Queen of Scots passed her weary
imprisonment; Aske Hall, near Richmond, a seat of the
Marquis of Zetland, the home of the leader in the
"Pilgrimage of Grace"; the Tudor rectory at Wath;
and, close to Wordsworth's "Hartleap Well," the
Elizabethan Walburn Hall.   Norton Conyers, near the

Marske Hall

little village of Wath, was the home of the Nortons, of
whom Wordsworth, referring to the 1569 rising, says in
"The White Doe of Rylstone,"

> Thee, Norton, with thine eight good sons,
> They doomed to die, alas, for ruth!

Some even of the more modern buildings that replaced
the fortresses of turbulent days have disappeared and

left little trace. One such was the grand Jacobean mansion that took the place of Malton Castle. The heiresses of its builder quarrelled about the ownership, and the matter was brought before the High Sheriff of Yorkshire, who, after the manner of Solomon, ordered the manor to be demolished and its materials divided between the claimants.

22. Communications—Past and Present.

The earliest roads in our county were no doubt the trails followed by our remote ancestors through the forest or over the moorland; in part, no doubt, the same tracks that the wild animals, instinctively choosing the easiest path, had made for themselves. The best and most plainly marked were probably not better than the hardly discernible footpaths through the ling which even to this day enable the traveller on the moors to reach his journey's end. No one would wander far from his birth-place unless urged by imperative reasons.

At a much later time, as we may read in Macaulay's *History*, it was a matter of no small danger and privation to travel along the highways. Not only were the roads deplorably bad but they were frequented by robbers like Dick Turpin, of whose exploits on "the road to York" so many tales are told. Less than a hundred years ago a journey from York to London was a serious undertaking, not, as it may be in a Scotch express, a pleasant interlude between lunch and tea.

North Yorkshire has, however, long been better off in its roads than most districts. Since the beginning of history it has been on the main line of communication. We have learnt a little about the military highways that the Romans made across the county; highways by which their legions marched to repel the attacks of barbarians from beyond the Wall. Expeditions from Scotland to England and from England to Scotland chose as a rule the east coast road along the flanks of the Pennines through the level Vale of York, rather than the passage across the mountains of Westmorland. And to-day the North-Eastern Railway is a link in the "East Coast Route," the line that provides the speediest communication between London and Edinburgh.

Even in their worst days the main roads of the riding have not been very bad. To-day they are exceedingly good. Leeming Lane, for instance, that part of the Great North Road from Boroughbridge northward through Leeming Bar, where tolls were collected, is as straight and level as any in England.

Nowadays the County Council has in its care the main roads linking the great centres of population. From among its members it appoints a Highways and Bridges Committee of which the duty is to have charge of all roads and bridges, and well the duty is performed. The excellent stone provides abundant road-metal, and the chief roads, those striking across the county from north to south and those traversing the two great dales on the west, are splendid. But in the country districts, where traffic is small and money for road-making is

scarce, the surface is often loose and stony. The view
of the grand ruins at Rievaulx or Byland is the reward
of a stiff walk. Intercourse across the ridge between
Swaledale and Wensleydale is restricted. The slight
depressions, like that over the Buttertubs, afford only
steep and rough mountain roads; even that across the
lower end of the dale, between Richmond and Leyburn,

Buttertubs Pass

tasks the cyclist or the walker. The block of moorland
in the east, intersected as it is by narrow steep-banked
dales, is particularly difficult to traverse. Even the
best road, that from Stokesley to Whitby by way of
Guisborough through Bilsdale, is decidedly trying, and
that which runs from Pickering alongside the railway,
though utilised by pilgrims to Lastingham and the

Killing Pits of Goathland, and the Bridestones on Egton Moor, is little better than a rough mountain road.

With the advent of motor traffic the roads are recovering much of their lost importance and some of the old coaching inns are now roused from their long sleep. But by far the most important means of communication is the railway. In some sense the North Riding is the birthplace of railways, for it was at Yarm that in 1820 the promoters of the Stockton and Darlington met, and Thomas Meynell of Yarm was chairman of the scheme. It is only right therefore that our county should be well endowed with this useful means of traffic.

Except for a few privately-owned lines, like that which conveys iron-ore from the Kilton mine to Brotton, the North-Eastern Company has the monopoly of rail communication in the riding. Its main line from York through Thirsk and Northallerton to Darlington across the Tees is reputed to be the finest permanent way in the world. Along it, over the level acres of the Vale of York, the expresses run at more than a mile a minute. The 1.9 p.m. from Darlington covers the $44\frac{1}{4}$ miles to York in 43 minutes (61·7 miles per hour). From the main line are thrown out spurs to the west and loops to the east. Westward goes a spur from Thirsk to Masham. Another from Northallerton follows the Ure up Wensleydale and climbs up the Pennine fells until on the bleak and lonely moorland at Hawes Junction— said to be the highest station in England—it meets the main Midland line. A third spur from Eryholme penetrates to Richmond at the gate of Swaledale, but the business to be expected in that secluded vale has not yet invited extension. The spur from Darlington through Barnard Castle and Bowes over Stainmoor is mainly in Durham.

One great loop to the east encircles in its fold the whole of the north-eastern moors. From York through Malton to Scarborough it runs close to the county

Muker Pass, near Askrigg

border. The coast line from the southern watering-place is a remarkable one, running high on the cliffs and exhibiting delightful coast scenes right on through Whitby to Saltburn. From Saltburn it reaches Middles‧brough and the business part of the county, and joins the main line again at Darlington. A minor branch through Pickering to Whitby runs through wild and

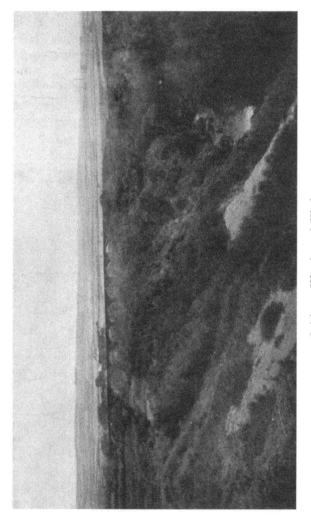

Saltburn Woods and Viaduct

rugged moorland, alternating with delightful stream and woodland scenery. Across the moors from Whitby the Esk has carved out a natural gradient, and railway and road here follow the river till they drop into the plain of Western Cleveland and reach the Tees-side towns. From Northallerton a line to the north-east meets the Tees-side branch at Yarm. In the industrial region we have a network of lines, showing how useful a hand-maid the railway is to the many industries that depend upon the mineral wealth of Cleveland.

The Ouse for some distance above York is navigable for barges, and the sluggish Foss has been canalised to the military camp at Strensall, the Aldershot of the north. The lowest reach of the Tees, too, has been converted by man's energy and skill into a deep ship canal. But of this we have already spoken in Section 9.

23. Administration and Divisions— Ancient and Modern.

Little is known of the system of government of the Brigantes, the early dwellers in our land, but we know a good deal about that of the Angles, who gradually formed their settlements here in Deira. The unit group or community was a "township," corresponding to the later parish. They had their dwellings within an enclosure, or *tun*, that protected them from the evil-disposed and their herds from wild beasts. Hence we have many town-names ending in *ton* or *don*—North-allerton and Malton, for instance, both places of importance in Roman times. The men of the township met at some conspicuous natural object, some big stone or isolated tree, in order to discuss their affairs and make their by-laws (laws of the *by* or township). All adult males might attend and take part in the deliberations, and it is a most interesting fact that some years ago a partial return was made to this primitive method, the settling of public questions in a meeting of all concerned. Parliament in 1894 established Parish Meetings for places with fewer than 300 inhabitants. To-day, therefore, all men in such places—in secluded villages like Lastingham or Old Byland—may meet in the schoolroom to manage local affairs, just as their forefathers met long ago in the open field.

The units were further grouped for convenience of governing into "hundreds" or "wards" or, in the North Riding, into "wapentakes." The hundred may have been an area originally settled by a hundred families, or simply a hundred hides of land—a hide being a portion of from 60 to 120 acres. Most likely, however, the name signified a district that was called upon in time of need to furnish a hundred men capable of bearing arms; for both the Scandinavian *wapentake* ("weapon-touching") and the Norman *ward* are military terms. They suggest that, when danger threatened, an organisation existed for defence. For the purpose of dealing with minor offences against the laws the North Riding is still divided into eleven wapentakes and the "liberties" (districts once governed by a great noble) of East and West Langbaurgh and Whitby Strand. Langbaurgh is the Cleveland area.

The meeting, or *moot*, of the whole shire was a body of great importance. It met twice a year for the transaction of business, usually to decide how the King's need for money was to be met. It was attended by twelve elected men from the wapentake and by four men with a *reeve* as leader from each township. Later, a King's officer, the reeve of the shire or *Sheriff*, came from the central authority to look after the King's interests.

The wisdom of the Conqueror allowed the old system of local government to remain, and it persisted long after the Conquest. It decayed as the power of Parliament grew and was called into existence again only at the beginning of the nineteenth century. Though similar to it in many ways the County Council is not the descendant of the Shire Moot. Our present local government is the creation of Parliament.

In a thousand years, however, the population, even in our sparsely-peopled county, has multiplied forty-fold, and government is no longer concerned simply with raising men and money for war and with the keeping of order. The work has become much more complicated. The moots of townships and wapentakes and shires are now represented by regularly constituted Parish Councils, District Councils, and County Councils. The County Council meets four times instead of twice a year; its various Committees are constantly at work; and it deals with the most diverse questions, with the health of the people, their education, and their well-being in various ways. The members of these governing bodies are elected by the people, so that we all have influence in the rule of our localities.

The astonishing growth of our large towns during the last two centuries has led to the creation of a new area of local government. Many towns have so great a number of inhabitants that a separate and independent council is needed in them. These large towns have, therefore, been made County Boroughs, which are independent and have all the powers of the Council of the County; the Lord Mayor or Mayor in each case performing the duties of the Sheriff. The North Riding has one County Borough, Middlesbrough, and in its affairs the County Council has no authority. Other towns there are, not quite populous enough to be created County Boroughs, but sufficiently important to have for most purposes their own governing body. The County Council has some authority over them though most questions are settled by the Mayor and his Town Council. The Municipal Boroughs, as these towns are called, in the North Riding are Scarborough, Thornaby-on-Tees, and Richmond. Other towns, like Northallerton, are the centres of Urban Districts, and the rest of the county is divided into Rural Districts.

The County Council holds its meetings in the County Hall at Northallerton. When it assembles in its quarterly meetings, it considers and discusses reports and authorises plans suggested by its Committees. These smaller groups, to one or more of which every Councillor belongs, are the active working bodies. They have frequent meetings and their members must give up a great deal of their time to promote the comfort and happiness of those whom they represent.

Victoria Square and Town Hall, Middlesbrough

The North Riding County Council contains sixty Councillors elected by the ratepayers, and twenty Aldermen (a third of that number) elected by the Councillors. The Councillors serve for three years, the Aldermen for six years; in other respects all have the same powers and duties.

Two other governing bodies must be noticed, the Guardians of the Poor, whose duty it is to see that the old and infirm do not suffer privation, and the great central authority, the Parliament at Westminster, of which the County Council itself is only the delegate.

The district attached to the parish church was once the unit for the care of the poor; but later, for the sake of economy, parishes were joined into larger areas called Poor Law Unions. There are twenty-three such unions in the riding.

Besides its peers who sit in the House of Lords, the North Riding sends six members to the House of Commons. Four of these are elected by divisions of the county outside the large towns of Middlesbrough and Scarborough, and these correspond to the old Knights of the Shire. The four county divisions are Richmond-shire, Thirsk and Malton (including strangely enough the Ainsty around York), Whitby, and Cleveland; the latter smallest in area but densest in population. Of the two borough members, corresponding to the burgesses first summoned by Simon de Montfort, one is elected by the voters of Middlesbrough, and one by those of Scarborough.

24. The Roll of Honour of the County.

One cannot indeed claim for our county some of the great barons that have been at one time or another associated with the riding. Bruce of Skelton Castle, where King Robert Bruce was born, the ancestor of the hero king of Scotland and so of our own royal house; the powerful Mowbray; the "King-maker" Warwick ruling

Wensleydale from his castle at Middleham, and Richard Crouchback who succeeded him there and was loved and regretted along the Ure; the Scropes of Bolton, who in an age of violence rose to eminence through their learning: these make only a passing show. They no more belong to the North Riding than the unlucky Queen of Scots because she was imprisoned in Bolton, or Richard II because once Pickering Castle held him. Men like Edmund Burke, too, who for many years was Member of Parliament for Malton before that ancient city was deprived of its members, have only an accidental connection with the riding.

Nor can we include in the Roll of Honour such men as the notorious highwayman Dick Turpin, one of whose exploits was his escape from Westmorland constables by leaping over Hell Gill into Yorkshire; or the still more notorious Duke of Buckingham of Charles II's reign. Buckingham, once lord of Helmsley, was among the worst of the dissolute courtiers around the pleasure-loving king. Deserted and ruined he died miserably at Kirby Moorside, "in the worst inn's worst room" says Pope, where in the parish register we may read: "1687, April 17th, George Vilaus, Lord dooke of bookingham."

The North Riding is glad, however, to claim a share in men like Wyclif and Coverdale, to whom religion owes much; in men of letters like Caedmon and Wordsworth; in men of action like Captain Cook; in men of inventive genius like Sir Lowthian Bell; even in men of great business capacity like Thomas Meynell and the "Middlesbrough owners."

Whitby is as proud of Caedmon as of its abbey. The Venerable Bede, who often visited Lastingham and Whitby, tells his story. He was a lay brother, an aged uneducated peasant, who tended the horses and oxen belonging to Whitby Abbey; and no man seemed less likely to become a poet. When at the feasting the custom was observed that all should sing in their turn he would depart to his lodging when the harp approached him. Having seen to his cattle one evening he slept and dreamt. One stood by his bed and hailed him by name, "Caedmon, sing me something." Then Caedmon declared his inability. "Yet thou canst sing me something." "What shall I sing?" said he. "Sing for me the beginning of creatures." Then the unlettered man was enabled to make poetry, and on the morrow Abbess Hilda took him into the monastery where he spent the rest of his life in learning and in composing poems. Here are the first four lines of his song; it will be noticed that they are not rhymed, but that each has two or more words beginning with the same letter:

> Nu we sculan herian/heofonrices Weard,
> Metodes mihte/and his modgethonc,
> wera Wuldorfaeder;/swa he wundra gehwaes,
> ece Dryhten,/ord onstealde.

(Now ought we to praise the guardian of heaven, the might of the Creator and his wisdom, the wisdom of the glorious father of men, and tell how he, the eternal Lord, appointed the beginning of every wonder.)

William Wordsworth and his sister spent much time in the North Riding: on the road between Leyburn and

John Wyclif

Richmond is the broken well about which he wrote the pathetic "Hart Leap Well," and his sonnet about the Hambleton Hills is a fine one. He was married in Brompton (Sawdon) church in 1802. Sir Walter Scott, too, loved the county: the "Rokeby County" was thoroughly known by him; Whitby and Abbess Hilda appear in *Marmion*, and Jervaulx is the Abbey of Prior Aylmer in *Ivanhoe*. Dickens visited Bowes, where he found the original of Mr Squeers in *Nicholas Nickleby*. For twenty years from 1809 Sydney Smith, "wittiest of Englishmen," was rector at the little village of Foston, and Laurence Sterne has made Coxwold and Shandy Hall famous. The great painter Turner delighted in North Yorkshire scenery and his pictures of Aysgarth Force, of Richmond, and of Rievaulx are among his best. At Pickering was born Francis Nicholson (1753–1844), one of the founders of the Water Colour Society in 1804, who, though never attaining the height of Turner or Girtin, was happy in depicting the river scenery of his county.

John Wyclif, "the morning star of the Reformation," earliest and in some ways greatest of the reformers, was born about 1325 in Wycliffe, the pretty village by the tree-shaded bank of the upper Tees. Miles Coverdale (1488–1568), sprang from the North Riding. He was of Coverdale, the secluded and delightful valley of the Cover that runs from Middleham up to Whernside. To him and his Bible of 1533 good judges attribute much of the beauty of the authorised version of 1611. Roger Ascham (1515–1568) the famous author of *The Scolemaster*, the friend and tutor of Queen Elizabeth and Lady Jane

Captain Cook

Grey, was born at Kirkby Wiske, where his father was in the service of the Scropes of Bolton.

In the evil days of Charles II, George Fox, the sturdy Quaker, taught in the North Riding; and visitors to Scarborough Castle are shown where, attacking those in high places, he was imprisoned in a room open to the sea "so that the water came over my bed and ran about the room, that I was fain to skim it up with a platter." His followers were once many in the dales and his methods of plain speech were, and are, imitated all over the riding.

On Easby Moor in the Cleveland Hills is erected a fine monument to Captain James Cook. The Yorkshire folk justly honour this great man, for to him more than to any other person we owe our possession of Australia and New Zealand. He embodied the best qualities of a Yorkshireman, was daring yet prudent, shrewd and hard working, skilful and enterprising, and he had what is so greatly valued in our county, any amount of grit. He belonged entirely to the North Riding, was born at Marton in 1728, had his schooling at Ayton ("Canny Yatton" the people call it), and was apprenticed to a worthy grocer at Staithes. When he left the distribution of groceries for a more strenuous life at sea it was from Whitby he sailed, and the abbey town provided him with the vessel in which he made his greatest voyage.

By no means to be omitted from our list are the modern captains of industry who have made the Middlesbrough district renowned. Prominent among them were Messrs Bolckow and Vaughan, to whom is mainly due not only the working of the Cleveland iron-seam but also the improving of the furnaces so that iron production is greatly cheapening. To them also is due the application of the studies of Gilchrist and Thomas, who showed that the Cleveland iron, though containing phosphorus, could be made into excellent steel. Here, too, Sir Lowthian Bell devised improve-

Sir Lowthian Bell

ments in iron production that have been adopted everywhere, though most of his experiments were conducted at Clarence on the other side of the river. Mr Thomas Meynell of Yarm, the first chairman of the first railway company, must be accorded a share in Middlesbrough's greatness; so also, of course, must Joseph Pease and his associates, the far-sighted and enterprising Quaker syndicate that cherished the infancy of the new town.

25. THE CHIEF TOWNS AND VILLAGES OF THE NORTH RIDING.

(The figures in brackets after each name state the population of the place according to the Census of 1911. Those at the end of each paragraph refer to the pages of this book in which the places are mentioned.)

Ainderby Steeple (235), on the Swale, 3 m. W. of Northallerton, is a pleasant village where Wensleydale merges into the plain. It takes its name from its conspicuous church spire. The chancel of the church is fine Decorated work similar to that at Patrick Brompton. (pp. 18, 51.)

Alne (441), a village 6 m. E. of Boroughbridge and west of the Forest of Galtres has some notable Norman work in its church, partly built of brick.

Ampleforth (701) is a stone-built farming village 4 m. S.W. of Helmsley on the slope of the Hambleton Hills towards the Vale of York. Near is the Benedictine Abbey with its college, the most important Roman Catholic establishment in Yorkshire. Byland Abbey and the camp of Studford Ring are not far off.

Askrigg (470), a pretty town in Upper Wensleydale. It was once famous for clock-making and for knitting, but these industries are now ousted by machinery. Two old Carolean houses face the bull-ring and the church, which is of Perpendicular date. (p. 129.)

Aysgarth (279, but in 8103 acres of the huge ecclesiastical parish, the largest in England, which stretches for miles to the borders of Westmorland, 1229) on the Ure in Upper Wensleydale. Its old woollen mill worked by the falls is now turned into a flour mill. Its church, which has splendid Perpendicular screen-work from ruined Jervaulx Abbey, looks over the bridge and the

Askrigg Hall

upper falls; the larger falls are some distance below the bridge. (pp. 4, 24, 28, 30, 123, 140.)

Ayton, Great (2319), 5 m. S.W. of Guisborough on the Leven overlooked by Roseberry Topping, is sharing the prosperity of the Middlesbrough industrial area. East and West Aytons (341) are small villages on the Derwent in Forge Valley about 4 m. S.W. of Scarborough. (pp. 115, 142.)

Bainbridge (587), a pretty dale village close to Askrigg, was the Roman station *Virosidum*. The Roman camp on Brough Hill, the road across the fells to Ingleton, the village green and the stocks are the chief things of interest.   (pp. 28, 59, 98, 100.)

Bedale (1163), a brick-built market town on a little feeder of the Ure at the foot of Wensleydale 7 m. S.W. of Northallerton, has a fine Early English church, with later additions.   (pp. 113, 121, 122.)

Boosbeck (5928), a busy ironstone mining town in the Cleveland ironstone district 3 m. S. of Saltburn.   The railway from Saltburn to Guisborough passes through it.   (pp. 71, 77.)

Bowes (577) the site of Roman *Lavatrae*, is a bleak village 4 m. S.W. of Barnard Castle on the Roman road and the modern railway over Stainmoor into Westmorland.   Here was "Dotheboys Hall" of Dickens's *Nicholas Nickleby*.   Its squat Norman castle is a good example of an early keep.   (pp. 93, 98, 100, 114, 129, 140.)

Brompton (1182) 7 m. S.W. of Scarborough, is a farming village on the road and railway from that watering place to Pickering.   (p. 140.)

Brompton-in-Allertonshire (1487), a farming village on the Northallerton and Yarm branch of the railway 1½ m. N. of the former town, is famous for its Saxon "hog-backs" (three perfect and fragments of ten others), perhaps the finest in England, and two Saxon crosses.   (pp. 74, 100.)

Brotton (3703 in township, 6780 in parish), an iron-mining and iron-smelting town in the Cleveland industrial region on the coast railway 2 m. S.E. of Saltburn.

Castleton with **Danby** (1164), a pretty farming village on the upper Esk near the edge of the Cleveland moors 6 m. S.E. of Guisborough.   It was of Danby, in which parish Castleton stands, that Canon Atkinson wrote *Forty Years in a Moorland Parish*. (p. 97.)

Bedale

Catterick (534), 4½ m. S.E. of Richmond, was a Roman station (*Cataractonium*), and an important post on the Watling Street, and is now waking to life again through motor traffic. Its fine church is wholly Perpendicular. (pp. 99, 100, 123.)

Coatham (4744), adjoining Redcar, with which it makes a single town, is really a seaside suburb of Middlesbrough, and is growing with the growth of the industrial region. On Coatham marshes are ancient salt "wyches," once worked by the Abbot of Guisborough. (p. 48.)

Cotherston or **Cotherstone**, a little village beautifully placed near the junction of the Balder and the Tees 3½ m. N.W. of Barnard Castle, is famous for its cheeses, and was formerly for its Quakers. (pp. 26, 93.)

Coxwold (294), 7½ m. S.E. of Thirsk, "in a rich valley under Hambleton Hills" says Sterne, was long the home of this witty writer who was curate here, and here at Shandy Hall he wrote *Tristram Shandy*. Byland Abbey is in this parish. (p. 140.)

Crayke (412), 2 m. E. of Easingwold on the western edge of the Howardian Hills, was till 1844 part of the domain of the Prince-bishops of Durham. (p. 14.)

Croft (494), where the main North Eastern line crosses the Tees into Durham, now attracts visitors by its mineral spring. Its interesting church possesses Norman, Early English, and Decorated work. Croft Bridge was the point where the Bishops of Durham entered on their possessions.

Cropton (338), 4 m. N.W. of Pickering on the southern outliers of the north-eastern moors, is near Cawthorne Camp, the most complete remains of a Roman fortification in the country.

Easingwold (2055), a sleepy town in the farming region north of Galtres Forest, stands on the high road between York and Northallerton. The bull-ring in the market place, the ruined cross, and an old half-timbered house are objects of interest.

Eryholme (168), on the Tees, 4½ m. S.E. of Darlington. Its red sandstone church has Norman and Early English work. (p. 128.)

Eston, with which is included in one Urban District **Grangetown** (12,026), at the foot of the Cleveland Hills is a mining appendage of Middlesbrough. The main ironstone seam is worked here, and here are the great steel-works of Bolckow, Vaughan & Co. Though a typical industrial town its situation makes it attractive. (pp. 19, 48, 71, 77.)

Gilling (714), on Gilling Beck in the Vale of Mowbray 3 m. north of Richmond, is a farming village that gave its name to the wapentake. It was probably superseded as the capital of the district by the Norman choice of Richmond.

Grosmont (918), a beautiful place where the Esk crosses the Cleveland Dyke above Whitby some 5 m. S.W. of that town. The dismantled furnaces and the slag heaps still show that the iron industry failed to establish itself here. (pp. 27, 38.)

Guisborough (7061 in the Urban District), a picturesque market town in the plain of the Leven under Roseberry Topping, 9 m. S.E. of Middlesbrough. The iron mines near are deserted for more productive ones. The chief pride of Guisborough is the west front, glorious in its ruin, of the Early English Priory, rising above the streets. A fine gatehouse abuts on the main street. (pp. 60, 76, 77, 103, 109, 127.)

Hawes (1518), a pleasant stone-built town on the infant Ure, is the market-place for Upper Wensleydale. Hardraw Force is a mile away, and Buttertubs Pass leads over the ridge to Muker in Swaledale. (pp. 18, 23, 27, 58, 62, 75, 128.)

Helmsley (1393), a pretty stone-built town on the Rye, not far from Rievaulx Abbey. Helmsley Castle ruins, inside Duncombe Park, was once owned by the Duke of Buckingham satirised by Dryden and Pope. (pp. 16, 29, 94, 107, 114, 137.)

Hinderwell (2491 with **Runswick** and **Staithes**), a village on the coast 10 m. N.W. of Whitby. (pp. 32, 36, 44, 45, 47, 48, 49, 50, 62, 80, 81, 82.)

Huntington (1326) is a quiet old farming village on the Foss 3 m. northward of York.

Kirby Moorside (1657, including **Keldholme**, where there is a Cistercian Nunnery), a stone-built, red-tiled town, stands on the Dove 6½ m. W.N.W. of Pickering. The Saxon shrine of Lastingham is conveniently reached from here. (p. 115.)

Kirkdale (48), a tiny village a mile W. of Kirkby Moorside is famous for its Saxon Church with an inscribed sun-dial over the south door and several Saxon crosses. Kirkdale cave, unearthed in a limestone quarry near the village, is as famous for its fossils as the church for its dial. (pp. 40, 101, 111.)

Lastingham (135), a lonely village on the north-east moors 6 m. N.W. of Pickering, had a famous Saxon monastery which Bede, who tells its story, often visited. Lastingham Church has a notable Norman crypt. (pp. 111, 127, 132, 138.)

Leyburn (832), the capital of lower Wensleydale, is prettily placed above the Ure on the limestone platform called Leyburn Shawl, which affords magnificent views of the dale. (pp. 18, 43, 127, 138.)

Loftus (5105), on the coast railway 4 m. S.E. of Saltburn, is an ancient town developing through its iron mines into an important industrial centre.   (pp. 16, 45, 75, 77.)

Malton, New (4822), on the Derwent, midway between York and Scarborough, was the site of a Roman camp.   It is the business centre for the southern part of the Vale of Pickering, and from its large race-horse stables is called the " Newmarket

Leyburn Ulshaw Bridge

of the North."   Malton as an ancient borough sent two members to the " Model Parliament " of 1295.   Edward Burke represented it from 1780 to 1794.   Castle Howard lies 5 miles to the eastward. **Old Malton**, a mile off, has as its parish church a beautiful Gilbertine Priory (date about 1150).   (pp. 14, 29, 30, 68, 123, 125, 129, 132, 137.)

Marske (2955) near Redcar, like it is growing with the growth of Middlesbrough.   Iron-mining and catering for the increasing number of visitors are the chief occupations.   Marske Hall (Earl of Zetland) is a beautiful Carolean house.

Marske near Richmond, is on a beck of the same name near its junction with the Swale.

Marton (1239) is an ancient village now becoming a pretty residential appendage to Middlesbrough. Captain Cook was born here and his monument stands on Easby Moor 6 m. to the S.E. There is another **Marton** on the Rye 4½ m. S.W. of Pickering. (p. 142.)

Market Place, Masham

Masham (1039), market town on the Ure at the point where Wensleydale merges into the Vale of York 8 m. N.W. of Ripon. A quiet country town, built round a large green, with an immense autumn sheep fair. Its church has much Norman work. (pp. 18, 28, 128.)

Middleham (680), south of Leyburn in Wensleydale, is an interesting stone-built village near the Ure. The Roman road and the "King-maker's" castle are notable. (pp. 28, 68, 101, 105, 114, 116, 137.)

Middlesbrough, now including **North Ormesby,** which in 1911 had a population of 14,582 (104,767), seven miles from the sea on the south bank of the Tees, which is here a great ship canal. Though of recent growth—a growth unparalleled in England—it is now one of the foremost cities in the world, contains a quarter of the population of the riding, and conducts almost all its foreign trade. Its iron and steel output is enormous and is increasing. It is the sixth port and the chief iron town of the country. A daughter of Darlington—it was called " Port Darlington " in earlier days before a syndicate of Friends from the Quaker town bought its site—it has far outstripped its mother. A tablet on the wall states that a house in West Street was built April 1830, "being the first house in the new town of Middles-brough-upon-Tees." It is a County Borough and returns one member to the House of Commons. (pp. 7, 12, 26, 45, 47, 48, 51, 53, 57, 62, 63, 66, 70, 71, 74, 78, 84–87, 94, 100, 121, 129, 134, 136, 142.)

Normanby and **Southbank** (14,877), an industrial town 3½ miles south-east of Middlesbrough, and like it dependent on the iron trade. It also makes bricks and glass. Together with Eston, with which it is continuous, it stands near the main iron seam at the foot of the Cleveland Hills. (pp. 26, 71, 84.)

Northallerton (4806), in the middle of the plain near the sluggish Wiske, is the official capital of the riding, having the county offices and prison. Of great historic interest, it was a Roman station, near a Roman road, and once stood in a kind of dependence upon the Bishop of Durham, who built here a castle and a palace. The Battle of the Standard (1138) was fought near the town. It has a large linoleum factory, makes leather and motor cars, and shows signs of development under railway influence, for from the main North Eastern line branches go from Northallerton east to Yarm and west to Hawes. There is a large trade in agricultural produce. (pp. 1, 6, 9, 60, 74, 92, 98, 99, 109, 111, 115, 128, 131, 132, 134.)

10—5

Patrick Brompton (135), a pleasant village in Wensley-dale on the road between Bedale and Leyburn has a beautiful Decorated church. (p. 111.)

Pickering (3674), an ancient market town that has given its name to the plain south of the eastern moors, and is the chief town of the Vale. The nave of its Norman and Early English church is decorated with a wonderful series of frescoes. Pickering Castle ruins are those of a castle of Norman masonry upon an English earthwork. The town is a favourite tourist centre. (pp. 4, 19, 43, 59, 68, 70, 79, 111, 119, 127, 129, 137.)

Redcar (5744), near the Tees mouth, a popular seaside resort for the Middlesbrough district, has a glorious stretch of sands. The iron industry is rapidly developing—there are already four blast furnaces at West Coatham—and it is a fishing station, the fishing village forming a strange contrast with the handsome buildings of the watering-place adjacent. (pp. 10, 31, 36, 44, 47, 48, 50, 63, 71, 79, 82.)

Richmond (3934), one of the most picturesque towns in north England, stands under its Norman castle built on a mound round which the Swale runs. Placed at the gate of Swaledale it also dominates the Vale of Mowbray. There is a good Perpendicular tower of the Grey Friars. Henry VII, to whom the earldom of Richmond ultimately came, bestowed the name on Sheen in Surrey; but the "Sweet Lass of Richmond Hill" was Yorkshire born. Richmond is a Municipal Borough. (pp. 4, 14, 29, 35, 63, 68, 70, 76, 93, 105, 109, 114–16, 124, 127, 128, 134, 140.)

Saltburn-by-the-Sea (3322) is a modern and very beautiful watering-place 4 m. S.E. of Redcar, where the Skelton Beck cuts its way through a deep glen. Some fine Italian gardens add to its attractions. Here begins the great cliff wall for which the Yorkshire coast is notable. (pp. 36, 48, 51, 64, 100, 129.)

Scarborough (37,201), perhaps the most beautiful of the larger seaside resorts of England, clusters round two bays separated by the huge buttress on which the castle stands, but joined by the new marine drive. It was important long before it became a fashionable watering-place. In 1066 Harold Hardrada burnt the city. Its castle has been a royal fortress since surrendered to Henry II. Its Spa waters, confirmed in fashion by Sheridan's *A Trip to Scarborough*, are only one among a great number of attractions—including a very interesting museum. Scarborough is a Municipal Borough, and returns one member to the House of Commons. (pp. 7, 8, 22, 29, 36, 38, 45, 47, 49, 50, 57–60, 79–81, 83, 87–90, 92, 94, 109, 114–16, 119, 134, 136, 142.)

Skelton (292), a little village 3 m. N.W. of York close to the Ouse, has a beautiful little Early English church, remarkable for the purity of its style. (p. 103.)

Skelton-in-Cleveland (8949), near Skelton Beck, 3 m. south-west cf Saltburn, is an important seat of iron-mining, and iron and steel working. Skelton Castle was formerly the seat of the great Bruce family. (pp. 48, 77, 91, 114, 136.)

Sleights (1353 with **Ugglebarnby**), near the Esk on the Whitby and Pickering road. On Sleights Moor are the " Bridestones," some curious megalithic remains.

Southbank (14,977), 3 m. E. of Middlesbrough, has large iron and steel-works. (pp. 26, 71, 84.)

Stokesley (1624), an old market town in Leven Vale 9 m. S.E. of Stockton, just outside the influence of the industrial region of Cleveland. The Leven runs through it, and a road across the moors through Bilsdale joins it with Helmsley. (pp. 16, 26, 127.)

Thirsk (2937), an old market town on Cod Beck—a little feeder of the Swale—in the fertile plain below the Hambleton Hills, 22 m. N.W. of York, has been rather unjustly treated by the railway. The main line passes a mile to the west, and the modern **Thirsk Junction** seems destined to outstrip in importance the mother town. The suburb **Sowerby** (2201), too, appears more living than the ancient town, Thirsk. The parish church is perhaps the finest Perpendicular church in the riding. (pp. 14, 19, 63, 104, 111, 115, 128.)

Thornaby-on-Tees (18,503), formerly South Stockton, is largely a result of the extraordinary growth of the industrial region. It is a prosperous iron-working town with blast furnaces, iron-works and steel-works. It is a Municipal Borough. (pp. 26, 53, 71, 134.)

Thornton-le-Dale (1181), on Thornton Beck, a tributary of the Derwent, and three miles east of Pickering on the railway to Scarborough, is a quiet farming village in the Vale of Pickering.

Whitby (with its suburb, Ruswarp, 11,139), "betwixt the heather and the northern sea," stands at the mouth of the Esk at the seaward foot of the Cleveland moors. Famous for its ruined abbey and its cross, it is also a fishing station, and has some trade. It is of exceptional interest to the historian since it was the home of Abbess Hilda, of the poet Caedmon, and later of Captain Cook. Its pier museum is interesting, especially for its fossils, a Saxon comb with runic inscription, and bones of extinct animals from Kirkdale. (pp. 8, 9, 26, 32, 36, 40, 43, 45, 47, 48, 57, 63, 75–80, 82, 83, 87, 89, 94, 97, 103, 105, 122, 127, 129, 131, 138, 140, 142.)

Wilton-in-Cleveland (1092), near the foot of Eston Nab, 3 m. S. of Redcar, is in the iron-mining region of Cleveland. Its castle, once the home of the Bulmer family which suffered for its share in the Pilgrimage of Grace, is now a seat of the Lowthers. The modern building is on the ruins of the earlier one. There is a small village called Wilton 4 m. E. of Pickering.

Yarm (1617), an ancient Tees port 19 miles up the river. It claims to be the "birthplace of railways," for here the Stockton and Darlington scheme was initiated. Under the energetic guidance of the Tees Conservancy Commission, Yarm may have a future, but at present it is far surpassed by its modern rivals. (pp. 26, 74, 87, 88, 128, 131, 143.)

Whitby

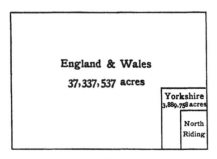

Fig. 1. Area of the North Riding (1,357,433 acres) compared
with that of Yorkshire and of England and Wales.

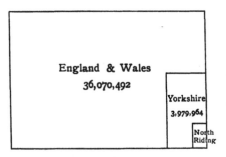

Fig. 2. Population of the North Riding (419,546) compared
with that of Yorkshire and of England and Wales in 1911.

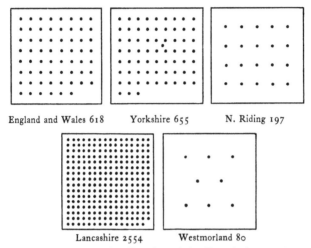

England and Wales 618   Yorkshire 655   N. Riding 197

Lancashire 2554   Westmorland 80

Fig. 3. Comparative Density of the Population to the
sq. m. in 1911.

(*Each dot represents 10 persons*)

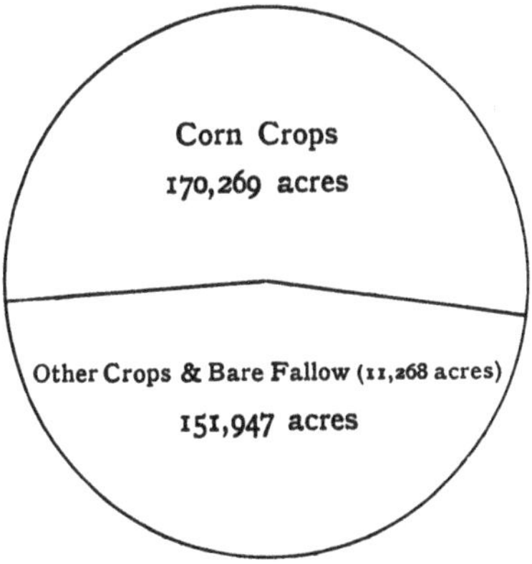

Corn Crops

170,269 acres

Other Crops & Bare Fallow (11,268 acres)

151,947 acres

Fig. 4. Area under Cereals compared with that of other
Farmed Land in the North Riding in 1914.

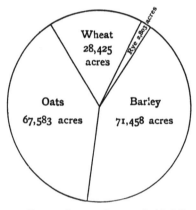

Fig. 5.   Proportionate Areas of chief Cereals
in the North Riding in 1914.

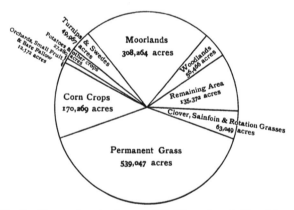

Fig. 6.   Proportionate Areas of Cultivated and Uncultivated
Land in the North Riding in 1914.

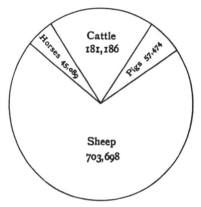

Fig. 7. Proportionate Numbers of Live Stock
in the North Riding in 1914.

The figures given in the several diagrams are retained, and
it will be noted that the conditions indicated are pre-war. The
proportions shown above depend mainly upon natural conditions;
these proportions, therefore, are likely to be restored when things
have settled down again.

Ingram Content Group UK Ltd.
Milton Keynes UK
UKHW010752250623
423921UK00013B/126

9 781107 622449